Flying on the Wings of Genius

A Chronicle of Modern Physics

Book 2

If I have seen farther, it is by standing on the shoulders of giants.

Isaac Newton (1676)

Dr. Andrew Worsley

Universal Publishers
Boca Raton, Florida

Flying on the Wings of Genius: A Chronicle of Modern Physics
Book 2

Copyright © 2006 Andrew Worsley
All rights reserved. No unauthorized copying of text, material, original equations or reformulation of original equations is permitted.

Universal Publishers
Boca Raton, Florida • USA
2006

ISBN: 1-58112-938-6

www.universal-publishers.com

About the Author:
Dr. Andrew Worsley,
(*nee:* Andrzej Wojciechowski)
Honorary Senior Lecturer, University of London.
Consultant, University Hospital Lewisham, University of London.
Author of: *On the Wings of Genius.*

Cover, *Angel of Wisdom* by: John and Bridget Wren Potter and Peter Ingram. Adapted form an original by Gustave Doré (1832-83)

Back cover by: Gemma Worsley.

Dedicated to:

The Elegance and Symmetry of Nature

The study of Nature is intercourse with the Highest Mind.

Jean Louis Agassiz (1807-1873)

Acknowledgements

This work is supported by the WWK Trust, whose aims are to promote worldwide ecological, scientific and cultural human development, by the sharing of knowledge.

We gratefully acknowledge the assistance of Dr. Nikolaos Mavromatos, King's College London, for his scientific support and discussion.

My gratitude also to my wife, Carolyn, a total non-physicist, and my brother Marek Koperski for their helpful comments and for making this book an enjoyable read.

Table of Contents

Forward..11

Chapter 1, Introduction.....................................15

Chapter2, The Standard Model........................19

Chapter 3, The New Quintessence 27

Chapter 4, The Harmonics of Matter................47

Chapter 5, The Music of the Spheres...............45

Chapter 6, The Spirit of the Atom....................53

Chapter 7, The Standard Model Remodelled..........59

Chapter 8, Vortex Harmonics..........................73

Chapter 9, Higher Order Quarks.....................81

Chapter 10, It must be Aesthetically Beautiful..........89

Chapter 11. Explaining the Neutrino....................99

Chapter 12, The final Piece of the Particle Puzzle....105

Glossary, From the Alpha to the Omega...............125

Technical Notes...131

References..135

Forward

We stand on the shores of a great ocean of truth examining pebbles on the beach..........while the whole sea of unfound knowledge lies before us.

Isaac Newton

Newton's original laws of motion held sway in the scientific world for over 250 years and some still apply today. What is scientifically now known is certainly more advanced than it was in the time of Newton. Yes we have invented computers, aeroplanes, space travel and the electric kettle. But if we are to be honest, the fundamental knowledge of the workings of the Universe still eludes modern physics. Some believe that they are on the brink of having a comprehensive theory of everything in the Universe. That theory is known as string theory. Nevertheless, there are many problems with string theory and it seems to predict very little from first principles. In truth the ocean of knowledge is vast and we have not even set sail upon the sea that lies before us. Until we do so, it cannot be said that humankind has found the full wisdom or maturity to progress its knowledge.

What knowledge we do have today on the workings of the Universe seems like a leaky boat. It is made from ill-fitting pieces and when it is put in the water it merely starts sinking. Without a proper ship modern science is confined to exist upon our island of knowledge, without ever knowing what lies beyond. Science does not know, what it does not know. It was the great physicist John Wheeler who once said:

We live on an island surrounded by ignorance. As our island of knowledge grows so does the shore of our ignorance.

Additionally, it would appear that our wisdom to use the knowledge that we do have has not advanced. Indeed the present use of our knowledge, such as nuclear technology, merely imperils our own existence. Many challenges lie ahead and at the current time it is clear that the very survival of humankind may rest upon advancing our knowledge and our ability then to use it wisely, in this new millennium.

To assist in this end, Book 1, of this series of three books, entitled "On The Wings of Genius" was written. That book was written to help gain an insight into the aesthetic structure of Universe and to give a far greater understanding of the true beauty in which Nature is designed. It contained the knowledge to build the ship to sail the oceans, so that we can discover more knowledge and perchance begin to use it wisely.

This second book, sets sail on a course, which is destined to reveal some of the most elegant secrets of the Universe. One of the most mysterious aspects of Nature is the way in which the fundamental materials (subatomic particles) of the Universe are so aesthetically constructed. This second book lifts the shrouds of mystery, which surround this subject. It does so in a way, which accurately agrees almost completely with currently held experimental evidence about the physical characteristics of these fundamental materials from textbook physics [1-3]

In it we also find that some of the fundamental constants of nature may be derived from first principles. This can be done without recourse to the simple contention that these are just fundamental constants, whose veracity cannot be questioned, but shows logically how these may be derived. It does so in a way that not only agrees with modern physics, but also explains the elegance and symmetry of Nature. It builds on the knowledge laid out in the previous book and does so in a way that strengthens enormously the truth of those findings. It opens a new window onto the physical Universe.

As with the previous book these findings have also been published in a scientific fashion and I again refer the reader to the first two references of the book.[4,5] This second book is comprehensive in itself, nevertheless it is highly recommended that the reader read Book 1 of the series in order to achieve an overall perspective of the power of this approach to physics. The previous book not only clarified many of the mysteries of quantum physics, but clarified the nature of energy.

This 2nd book further clarifies the nature of energy and overall brings an even greater understanding of the magnificent symmetry of the physical Universe and the unified nature and exquisite elegance of its design.

Andrew Worsley, February 2006

Chapter 1

Introduction

The more one chases quanta, the better they hide themselves.

Albert Einstein

It is now over 100 years ago that a physicist named Max Planck discovered the fundamental constant of quantum physics. [6,7] Planck's constant, as it is still known today appears in virtually every quantum formula. These formulae for quantum physics have over these past hundred years been built up to form an entirely new picture of the world. That is a world where everything, on a small enough scale, comes in discreet packages or quanta. Nevertheless, even today the true nature of the quantum has evaded science. The quanta that have been sought for so long are indeed incredibly important.

In present day physics we are very good at measuring things and examining the pebbles of knowledge we do have in great detail. The problem is in doing so we can miss the overall picture of what is quantum physics. The fundamental question, which is not answered is, Why? Take for instance the electron, which is the particle, which is best known for the transmission of electricity down an electric wire. We *do* know the exact value of the mass and electrical charge of the electron particle. But present day physics has not got a single clue as to why the mass and charge of the electron is what it is. Similarly with many aspects of physics, there are constants of nature to which we have measured the value down to an accuracy of 12 decimal places, that is like measuring the distance between here and New York down to the thickness of a single atom. But we still have no idea whatsoever why that value is the value that it is.

In truth the claims to this tremendous accuracy regarding these constants of nature may not be that well founded. Yes we have measured the constants on this planet to that degree of accuracy, but what of confounding factors, that we are aware of, or even more likely unaware of. One of the biggest surprises of modern physics (although some still contend it) is that one of the important constants of nature, the fine structure constant, does not appear to measure the same everywhere we look. Some distant galaxies seem to have a fine structure constant which is slightly different. This difference is only some 1 in 100,000 but this is still significant. Furthermore the fine structure constant is dependant on other constants, such as the charge of the electron amongst others. So we really do not know, which if not all of these component constants are accurate. Moreover, the fine structure constant is just about the only constant we can measure on distant galaxies, by examining the light that comes from them. We cannot for instance go to these distant galaxies and measure the charge of the electron, so it is quite possible that other constants could be a little bit different elsewhere. There is another parameter that scientists *can* measure in distant galaxies, that is the ratio of the mass of the proton to the mass of the electron. Indeed recently scientists have noted that there also appears to be difference in the ratio of these masses, on distant galaxies, by about the same amount as with the fine structure constant. Until this mystery is solved this means we can realistically only be certain of values down to about five decimal places.† This still represents a considerable degree of accuracy. There is a model of the particle world called the Standard Model, this does not predict the fundamental constants or the masses of the fundamental constituent particles. The predictions it does make about subatomic particle masses, such as the proton are accurate only to 2 decimal places. The magnificent strength of this new approach, over the Standard model, is that it is possible to define these constants and particles from first principles to a much greater degree of accuracy. In doing so we can also solve the mystery of why the constants and fundamental particle masses are what they are.

This can be done, by finding the true ephemeral nature of the quantum. In the first book of the series we *were* able to find such a fundamental quantum and indeed this enabled the introduction of a unified description of the Universe. In that book we were also able to clarify some of the mysteries about the nature of the quantum formulae and derive them on an entirely logical basis. This led us closer to an understanding of the truer nature of energy and matter. This second book goes farther than this, it enables an even more comprehensive understanding of energy and in particular matter at a much more elegant level.

---

† In this book therefore we will operate on the basis of 5 decimal place accuracy.

What then of string theory, which has been the major contender for a comprehensive physical theory, over the past twenty-five years.[8-10] In the previous book we solved many of the problems associated with string theory. Another major problem of string theory is that it also fails to predict any of the fundamental constants or particles. So much so, that physicists are beginning to be concerned by its lack of predictive power. This has resulted in one eminent physicist in string theory saying:

"people in string theory are very frustrated as am I by our inability to be more predictive after all these years."

David Gross (2006).

Yet another problem is that string theory *does* modify the established equations for quantum physics. [11,12]

The truth is, that only certain elements of string theory are correct, and the fundamental quantum mass that is used in string theory is incorrect. Once we find the one true elusive quantum mass, we are open to deriving not only the equations for quantum physics from first principles but also the fundamental constants and the masses of the particles in an entirely harmonious way.

Chapter 2

The Standard Model

How can it be that writing down a few simple and elegant formulae, like short poems governed by strict rules such as those of the sonnet.....can predict universal regularities of Nature

Murray Gell-Mann

In this book it will be possible to demonstrate that a few direct rules are all that are needed to explain the ocean of knowledge that lies before us. However, before we embark on a journey across this ocean of knowledge we should recap on the history of particle physics. The first evidence of a subatomic world came with the discovery of the electron by J.J. Thomson in 1897. He first described what was thought of as a particle with a negative charge and a tiny mass. His discovery was made, using a cathode ray tube, which in fact is pretty much the same as an old-fashioned cathode ray TV tube. The discovery was incredibly important, the applications of electricity and electronics or the understanding of chemistry is unthinkable without the discovery of electron. Moreover some radioactive decays which are crucial in the understanding of particle physics, called beta decay, is actually the release of an electron from the inside of an actual atom. But as with many truly crucial discoveries the great importance was not recognised at the time.

Of the electron he said:

"Could anything at first sight seem more impractical than a body which is so small that its mass is an insignificant fraction of the mass of an atom of hydrogen."

J.J. Thomson.

About the same time as J.J. Thomson was working on the electron a famous Polish scientist, Marie Curie Skłodowska, was working in collaboration with her Husband Pierre Curie, on radioactivity. She showed as early as 1897 that certain naturally occurring atoms, such as Uranium, would decay by giving off small particles, which were themselves the constituents of the larger atom. The achievements of Marie Curie Skłodowska are made even more remarkable in that she was largely a self-taught woman in a field, which was almost exclusively dominated by men. Nevertheless, her discoveries were so important that she is one of the few scientists, and the only woman to date, to have won two Nobel Prizes. Moreover, what knowledge she had gained on splitting the atom she used for good purposes and she was able to set up an institute for the treatment of cancer with the very radioactivity that she had discovered.

What she had in fact discovered, from a physics perspective, were the decay products of the centre of the atom. This centre or nucleus was later found to be made up of small particles known as protons and neutrons. As early as 1900 she wrote a paper on the release of alpha particles, "Les Rayon Alpha" as she called them. [13] It was not for some time after this that it was realised that an alpha particle was actually two protons and two neutrons bound together, and was an important part of the decay of the nucleus of an atom.

Indeed it was not till 1918, that Ernest Rutherford actually discovered the proton itself. The proton in contrast to the electron had a positive charge, which was the exact but opposite charge of the electron. By 1919 a picture of the atom was emerging of a small central nucleus, orbited by electrons. So in the case of hydrogen, the smallest atom, we have a picture of a single proton orbited by a single electron. The nucleus of the atom was so small that if the orbit of the electron is the size of the dome of St. Paul's Cathedral, then the nucleus would be about the size of a marble. The nucleus is effectively 100,000 times smaller than the atom itself.

When the neutron, a neutral particle with no net charge, was later discovered by Chadwick in 1932 the picture of the atom seemed complete. Each atom had a central nucleus, which was composed of protons, with neutrons to stabilize the nucleus. The nucleus was itself surrounded by a cloud of electrons, which orbited around the nucleus. Differing atoms would have increasing numbers of protons and in turn electrons and these in turn would determine the characteristics of an atom.

This gives the wonderful diversity of atoms, which enables the world to be built up. For instance if we take two atoms of hydrogen (each having one proton and electron) and add one atom of oxygen (having eight protons, eight neutrons and eight electrons), these atoms go to form the molecule H_2O; that substance which is so essential to life - water.

Similarly if we take six protons and six neutrons and six electrons, we get the carbon atom, the principle atom from which all life on this planet is based. In this way with differing numbers of protons and neutrons we can build up all the atomic elements. Additionally, if we combine these atoms together, as with water, we can account for all the molecules we see in the Universe

Everything appeared so neat and tidy. But almost as soon as scientists had confirmed the presence of these three fundamental particles, the proton the neutron and the electron, then in 1936 came along another particle called the muon. To most physicists of the time this was an unexpected and unwanted surprise.

Certainly what was to happen next, came as an even more uncomfortable surprise, soon more and more particles started to appear. These came from the study of cosmic rays, specifically rays that are coming from outer space. Something coming from outer space was telling us that our simplistic model of the Universe was actually wrong. It was almost as if the Universe was hinting to us that that we were missing something very important. After their experiments with cosmic rays, scientists of the 1950's went on to built particle accelerators, these effectively accelerated protons (or electrons) and smashed them together at great speeds. Soon a whole plethora of particles were being revealed. These came in pairs, which seemed to mirror each other. So in the case of the electron with a negative charge they later discovered it's opposite called the positron, with a positive charge. This plethora of subatomic particles thus became known colloquially among physicists as the "particle zoo".

Particle physics was in deep trouble, if all the members of the particle zoo were indeed fundamental particles then the Universe was awash with these apparently fundamental particles. The picture of nature was looking far too messy. The particle physics ship was sinking and fast.

Along to the rescue came the eminent physicist Murray Gell-Mann. He surmised that the fundamental particles like the proton must be composed of even more fundamental particles, which were termed quarks. In the first instance *three* quarks were postulated. In the case of the proton, for instance three of these quarks were to combine to from the proton particle. But then as the number of new particle discovered grew, the number of quarks grew to six. With these six fundamental particles physicists were able to devise a system, which apparently explained all the particles that were proton and neutron like.

The history of science can however be strange. In fact it was two people that postulated the theory that there may be very fundamental particles at the core of particle physics, virtually at the same time. Murray Gell-Mann and George Zweig almost simultaneously, in 1964, described these fundamental objects. George called them "aces" and postulated that there were four such particles. Murray called them quarks and postulated that there were three of these. Indeed he called them quarks form a line in James Joyce's poem, Finnegan's wake "three quarks for Muster Mark". Murray Gell-Mann was theoretically nearer the mark at the time with three quarks, and won the Nobel Prize in 1969. George Zweig for his pains was not rewarded and left physics to become a neurobiologist.

The genius of Murray's theory is that it predicted the existence of three quarks called: up, down and strange and at the same time was able to predict the existence of new particles. When such a predicted new particle containing three strange quarks actually turned up in experiments later in 1964, Murray's theory was vindicated. Moreover the theory of quarks has been built up to produce an enormously successful model of particle physics called the Standard Model, which predicted the existence of a further three quarks

The next set of three quarks to be later discovered were originally called: charm, beauty and truth. Sadly the latter two have since been renamed, bottom and top - leading to various papers which have titles that for instance refer to: "the discovery of the bottom". There are some who plea for their names to be returned to their original, beauty and truth.

Nevertheless the Standard Model has been remarkably good at predicting certain aspects of physics. But as time goes by there is a realisation that this model does suffer from some drawbacks. Of the masses of fundamental particles, in one of the latest books on "Particle Physics", by Abraham Seiden,[14] it is said that

"There is no fundamental understanding of the mass pattern."

and

"The large variation in masses is an unsolved puzzle."

But not only are the masses of the fundamental particles like the electron and quarks not understood, the calculations which result in the masses of the known particles resultant from the current quark model are not quite perfect.

"The agreement is, however, not perfect and typical of a phenomenologically motivated model; we do not know how to improve the agreement systematically."

The time has come to investigate new models of nature, one that can offer us a window into why the fundamental particle masses are what they are and can explain why the fundamental constants of nature such as the fine structure constant and charge of the electron are what they are. A predictive theory is required which can completely unify the whole of physics.

Chapter 3

The New Quintessence

The mathematical sciences particularly exhibit order, symmetry and limitation and these are the greatest forms of the beautiful

Aristotle (384-322 B.C.)

The ancient Greek philosophers were amazingly insightful given their low level of technical scientific knowledge. They coined the term atom, to mean the smallest indivisible part of matter. They worked out many mathematical formulae, including Pythagoras's theorem about the ratio of the length of the sides of a triangle. Additionally, Aristotle worked out that there were four states of matter. The four essences as they were known, earth, wind, water and fire do actually represent the four known states of matter, namely: solid, gas, liquid and plasma. He even conceived a fifth essence or quintessence, which was a pure substance of which the heavens were made

How right he nearly was. Here it will be shown that there is a fundamental quantum, which we will also call quintessence, but its role is far more important. It is not only the quanta of which the heavens are composed, but far more than that. Ultimately quantum theory points to a fundamental constituent not only of the heavens, but of all matter, the forces of nature *and* of space time itself. This is that ultimate quintessence, where everything in the physical Universe is made of exactly the same exquisitely ephemeral quantum.

Modern science has conceived of a theory, which encapsulates this all embracing concept and it is called string theory. [8-10] Sadly there is one major thing wrong, which spoils this theory, the quantum mass used is incorrect. Planck's constant is in effect a very tiny mount of energy (actually energy x time). The mass quantum, is based on a mass called the Planck mass, whose value is far too high. Matter is composed of subatomic particles but the Planck mass is much, much higher than the mass of the subatomic particles it seeks to explain. The theory is right in one way but it is looking at far too large a scale. However if we go down to a small enough scale then everything in the physical Universe is based on harmonics. Indeed we can explain the formation of matter from first principles, but only if we use a scale; far, far smaller that of the subatomic particles. These are the ephemeral quintessential quanta from which the subatomic particles and all else in the physical Universe are constituted.

These are not the same quintessence that the eminent physicist Lawrence Krauss has described to explain the continued expansion of the Universe. They not only explain the continued expansion of the Universe but they account for space-time itself, as well as matter and the forces of nature. All these different aspects of physics arise from the very fundamental fabric of the Universe, specifically from a fundamental quintessential quantum from which the Universe itself is made.

Such a fundamental quintessential mass quantum has already been defined in Book 1. From this fundamental quantum, a new approach to physics will be revealed, one so powerful that the constants of

Nature and fundamental particle masses may arise from first principles, on an entirely logical basis. We will first briefly recap on the structure of this fundamental quantum. Mathematically we can derive the mass quantum in two ways, both methods corroborate the validity of the other method for the derivation. The first method in a self-evident way uses Einstein's standard energy mass relation formula $E=mc^2$, to calculate a fundamental quantum mass. (see Box 1).

Box 1.
1. Quintessential Mass Quantum (m_q)[†]

As $E = mc^2$

$m = E/c^2$

substitute
E for h,
then $\qquad\qquad m_q = h/c^2 \qquad\qquad (1)$

2. Quintessential Mass Quanta (m_q)

m_q = Planck mass x Planck time

$m_q = \sqrt{(hc/G)} \times \sqrt{(hG/c^5)} = h/c^2 \qquad (2)$

where $\sqrt{}$ is the square root, h is Planck's constant, c is the speed of light and G the gravitational constant. *For dimensions, please see technical note 2.*

[†] Dimensionally m_q = [M.T] multiply by n, which is the number of quanta *per unit time*, with the dimensions of [T⁻¹] we get the dimensions [M.T] x [T⁻¹] = [M].

The second method for the calculation uses the standard Planck mass, multiplied by the standard Planck time. Both methods elegantly give the same value for the quantum mass. At the same time this mass value then puts the fundamental quantum mass on the same footing as Planck's constant itself. Both Planck's constant and the quantum mass are equivalent in size. Planck's constant has a time function as part of it's energy component. Now the standard mass quantum has a time function as part of its mass component.[†]

Once we have this parity between mass and energy in the quantum world, then the unification of physics, becomes not only feasible, but results in a far greater understanding of the fundamental constants of Nature.

So how small is this new ephemeral quintessence compared to the original Planck mass? Well if the old Planck mass would have the weight of an entire galaxy, that is equivalent to the mass of a billion solar systems, then this new quantum would be the equivalent to the weight of a pinhead. That is how small the true quintessence is compared to the original Planck mass. Amazingly the original Planck mass itself actually has only a mass of a grain of sand, so the new quintessence on the scale of human imagination is immeasurably small. But this is what is required to understand the true world of quantum physics. Suffice to say that if we take the smallest thing we can measure the mass of (which is the electron), divide it by a hundred billion, billion (100,000,000,000,000,000,000)

[†] For a fuller treatment of the quantum mass please see Book 1, Chapter 7 Quintessential mass quanta. Please also see technical notes 1 and 2.

and we would have arrived at the fundamental quantum of the Universe. How can we prove this? As it is, we can measure energy to a far greater accuracy than we can mass – and the smallest amount of energy we have is Planck's constant and this new quantum mass is equivalent to the smallest unit of energy we have. Now if you take the energy of an electron and divide it by Planck's constant then you get exactly the same ephemeral number of quanta (100,000,000,000,000,000,000), per unit time,[†] that is, it is one hundred billion, billion times smaller. Planck's constant is the basis of virtually all of quantum physics, hence it would be entirely logical to make the equivalent quantum mass the basis of quantum physics.

Having done this, we have shown in Book 1 that virtually all of the equations for quantum physics drop out from first principles. The next stage is to see if the constants of nature also drop out from first principles. To do this we first need to quantify what the wavelength of this quintessential quantum is. Lets first of all use the standard equation for the wavelength of light (see Box 2). Importantly this is also the same as the equation of the *standard* wavelength of matter,[1-3,15] so if by all accounts both matter and light are made from the same quintessential quantum, then the same equation should apply. Indeed if we take a single quintessential quantum mass (as defined in Box 1) and put it in the standard equation for the wavelength [15] we get what is a very interesting answer.

[††] The number of qunata n = the number of quanta *per unit time*, so it will have the same dimensions of frequency, specifically $[T^{-1}]$. See technical note 1 and 2.

Let's see what happens mathematically if we take for instance a single quantum of (non rest) mass of light, which itself is travelling at the speed of light c, (Box 2).

Box 2
Quintessence wavelength (λ_q)

As: $\lambda = h/p = h/mc$
For a single quintessence *quantum, with a frequency of 1 per unit time.*
As: $m_q = h/c^2$ *(Eq, 1), then*

$\lambda_{q1} = hc^2/hc$

† $\lambda_{q1} = c$

For a single quantum unit with a frequency of 1 per unit time,
In S.I. units

$\lambda_{q1} = c$ *(in metres)*

$\lambda_{q1} = 2.9979245 \times 10^8$ m

For a single quantum unit with a frequency of 1 per unit time,.
In cgs units

$\lambda_{q1} = c$ *(in centimetres)*

$\lambda_{q1} = 2.9979245 \times 10^{10}$ cm $= 2.9979245 \times 10^8$ m

$$\underline{\lambda_{q1} = 2.9979245 \times 10^8 \text{ m}}$$

---

† Dimensionally, $\lambda_q = [L.T^{-1}]$ *divide by* n, which is the number of quanta *per unit time,* with the dimensions of $[T^{-1}]$ we get the dimensions $[L.T^{-1}] \div [T^{-1}] = [L]$.

The fascinating thing is that for a single quintessence travelling at the speed of light, then the wavelength is itself equivalent to the distance travelled by the speed of light in a single unit of time, whatever units you use. In the above example (see Box 2) we use two standard units, Standard International (S.I.), and centimetres, grams and seconds (cgs), the answer comes out exactly the same in each case.

The first question one might ask is what proof do we have of this. Well there is already plenty of proof around. Let's take an every day thing like a radio station, which transmits radio waves at a specific frequency, say 105.6 MHz. In this case there are effectively 105.6 million quanta in this particular radio wave. If we calculate the wavelength based on the above principles then it has a wavelength of 2.83894 meters, if we measure the actual wavelength of such a wave we get exactly the same answer. There are literally millions and millions of examples in science (and everyday life) were this relationship holds true. To this end we base our unifying principle on this elegant musical relation to the speed of light.

This relationship between time and wavelength may be a surprise to some, for it is taught that units are arbitrary, but in reality they are not. Any units we use have actually got to be self consistent and linked by the formula $E= mc^2$, that is energy and mass and space-time are linked. Scientifically, as detailed in Book 1, recent evidence from cosmology tells us that there is energy inherent in empty space. This corroborates the fact that space-time and energy are intimately connected and this brings us to the paradigm shift, that energy-space-time are linked. This puts us one step further in our understanding of the workings of the Universe.

The equation $E = mc^2$, itself implies that there is a link between space-time and energy, but the quantum leap of thought is difficult to make. Now that science has found the presence of energy in empty space, the link becomes almost *de rigueur*.

But wait, I hear some of my learned readers say, would I get the same answer if I change my set of time units to minutes, for instance. Well the answer is no if you just use it to work out the wavelength. *But provided you relate the wavelength to something with mass or energy the result effectively comes out the same*. This is because if you change the time units then you have to change your units of energy, mass or length to match. Indeed if you do the calculation with any units you care to chose, provided everything is self consistent and you relate it to something with mass or energy, you still do get the same answer. *Quad est demonstrandum* (see Box 2.). No, no I still hear you say, this still implies that the second itself has some special importance. Indeed it has, precisely because seconds are linked to both length in meters *and* the respective mass in kilograms and in turn to the unit of energy we use (Joules) by the very equation $E = mc^2$. We may of course use any unit of time, but you would need to change the other units and you effectively get the same answer.[†]

How have we come across the concept of the unit of time – partly through the reality of music itself. Take the note middle C it has the *scientific* frequency of 256Hz., which is mathematically 2^8 Hz The C note in

---

[†] See technical note 1

the next octave is doubled to 512Hz, which is mathematically 2^9Hz. When the frequency is doubled the wavelength is then halved. The next three octaves take us to 2^{12} Hz and in each case the wavelength is halved again. The interval between each note is itself mathematically a factor of $2^{1/12}$ Hz. The study of music itself has had an incredibly important effect on our appreciation of time and gives us a clue to the symmetry of the Universe.

The physics of quantum mechanics is similar to that of music itself. Music theory is essentially where string theory emanates from. But the quantum used here is different from the quantum used in string theory, in that it is far, far smaller. Thus according to the standard equation (in Box 2), using this new quantum, every time you double the number of quanta, then you half the wavelength. As it turns out, the music of the Universe is far more elegant than suggested by string theory. As we will show, the constants of nature, such as the charge of the electron and the masses of the subatomic particles such as the quarks may fall out from first principles on the basis of the harmonics of the speed of light in an exquisite fashion.

Chapter 4

The Harmonics of Matter

It is my opinion that everything must be based on a single idea, once we have finally discovered it, it will be so compelling so beautiful that we will say to one another, yes, how could it have been any different.

John Wheeler.

In science it is important that a minimum number of precepts are used to describe physical reality. In current physics there are a plethora of differing observations, which have disparate equations and theorems to explain them.

One of the beauties of string theory is that you can begin to base everything in the Universe on a single mass, this is known as the Planck mass. However, when we use this mass, it is so large that it is very awkward and current string theory has great difficulty in predicting what we find in Nature. Moreover, there now appears to be a plethora of differing string theory solutions. For a theory to be truly compelling it cannot have a plethora of possible solutions and far more importantly it has to be predictive. Specifically it must explain what we see and measure around us. In Book 1, the fundamental quantum mass of the Universe was redefined in such a way that the equations for quantum mechanics could be predicted, and derived from a single principle. In contrast, string theory, by using a very large quantum mass, can violate the equations for quantum mechanics.[11,12] Moreover string theory fails to be predictive. Unlike string theory, in this book we go on to predict many more things using the new quantum principles. These are parameters which string theory has been unable to predict.

Let us commence with the two most fundamental particles of the Universe. Firstly the electron, this is the particle which delicately and swiftly orbits the centre or the nucleus of the atom. Secondly the proton, this is the particle that actually sits in the nucleus of the atom. The wavelength and mass of the electron itself will be derived first and then it will be shown that the proton is a very special harmonic of the electron.

First we take the ephemeral quintessential energy quantum of the Universe, As previously described its velocity is the velocity of light (c=2.99792458 x 10^8 m.sec^{-1}). The wavelength for a single unitary quintessence is the velocity of light, in meters †(λ_{q1} =2.99792458x10^8m, see Chapter 3, Box 2). In accordance with music theory as the harmonic goes up the wavelength goes down. For instance the number quanta for the wavelength to be one meter has to be equivalent to the number of meters covered in one second by light (2.99792458 x 10^8 sec^{-1}), and that is the number of harmonic quanta required for the

---

† Dimensionally, λ_q = [L.T^{-1}] divide by n, which is the number of quanta *per unit time,* with the dimensions of [T^{-1}] we get the dimensions [L. T^{-1}] ÷[T^{-1}] = [L].

wavelength of a subatomic object to have a wavelength of one meter.

Using the perfect geometry of the sphere we can then take things to the next level of understanding. The volume of the sphere is based on the standard equation ($V = 4/3\ \pi r^3$). The wavelength of each individual quintessence is equivalent to the length travelled by light in a unit of time, i.e. it is dependant on the speed of light (see Box 2). So substitute the radius r for the term for the speed of light c, and we have true magic. If we take the reciprocal of the volume of such a sphere we go into the subatomic realm. As a result just about everything physical then may arise from first principles.

In the later chapter entitled "the Music of the Spheres", it is shown that the charge of the electron depends on the square root of such a light sphere. Additionally we show that a very important constant known as the fine structure constant depends on the square of such a light sphere. That is just the start, almost everything else may also fall out from the most logical of principles.

Here these principles are introduced mathematically by stating that if we take the square root of such a light sphere, we can readily work out the standard wavelength of the electron, known otherwise as the Compton wavelength of the electron.

Just to recap, we showed in Book 1 and here, that the frequency is dependant on the number of quanta contained within a single subatomic particle. Such that :

$$f = n \qquad (eq.\ 1)^\dagger$$

So in effect the frequency is determined by the number of quanta contained within a particle per unit time.† So using the principle that *c* numbers of harmonic quanta have the wavelength of 1 meter, then by the same token we need to increase the number of quanta by further powers of *c*, to get the wavelength of fractions of a meter so that we get down to a subatomic level. Specifically if you want to get a wavelength of $1/c^{1½}$ in meters, then you would need to divide by $c^{2½}$, this being the number of harmonic quintessence quanta required for that wavelength. As the frequency and the wavelength are directly related to the velocity, in this case the velocity of light, we should find that the number of quanta required to make up the electron and in turn the wavelength of the electron also depend on the velocity of light (see Box 3).

Now all we need to do is take the wavelength of each quantum (λ_q) and divide it by the number of quanta contained within any particle in this case the electron and you immediately get what in standard physics is known as the Compton Wavelength. This is what happens in music theory you take the wavelength and divide by any increase frequency you get a shorter wavelength. Now the Compton wavelength is

---

† The number of qunata n = the number of quanta *per unit time*, so it will have the same dimensions of frequency, specifically [T^{-1}]. See technical note 1 and 2.

† The number of qunata n = the number of quanta *per unit time*, so it will have the same dimensions of frequency, specifically [T^{-1}]. See technical note 1 and 2.

something that can be readily measured experimentally, when applied to the electron. So we will begin by addressing the Compton Wavelength of the electron.

Box 3,
Compton wavelength of the Electron (λ_{Ce})[†]

$(1/c^3)^{1/2} = 1/c^{1\frac{1}{2}}$ and $1/c^{1\frac{1}{2}} \times 1/c = 1/c^{2\frac{1}{2}}$

and taking c as the number of harmonic quanta/sec

$$\lambda_{Ce} = 2 * \lambda_q / c^{2\frac{1}{2}} = 3.862 \times 10^{-13} \text{ m}$$

Actual Compton Wavelength (λ_C)

$$\lambda_{Ce} = hc/2\pi m_e = 3.862 \times 10^{-13} \text{ m}$$

where c is the speed of light, ε is the standard electron magnetic moment to Bohr magneton ratio, and * = $(1 + 2\varepsilon)$, h is Planck's constant and m_e is the mass of the electron.

Q.E.D. The quantum wavelength of the electron - from the speed of light.

Working on these same principles gives an awesome insight in what is actually going on in the subatomic world. Not only can we get the wavelength of the

---

[†] Dimensionally, λ_q = [L.T⁻¹] divide by n, which is the number of quanta *per unit time,* with the dimensions of [T⁻¹] we get the dimensions [L. T⁻¹] ÷ [T⁻¹] = [L].

electron we get its mass and the same goes for the proton and all the known subatomic particles – and the effective mass of the quarks. All this and more comes out of the speed of light.

In order to derive the total mass of a system you need to take the fundamental quantum mass (see Box1) and multiply it by the number of those mass quanta. So using the same principle that c numbers of harmonic quanta have the wavelength of 1 meter, then by the same token we need to increase the number of quanta by further powers of c, to get wavelength of fractions of a meter so that we get down to a subatomic level. Specifically if you want to get a wavelength of $1/c^{1/2}$ in meters, then you would need $c^{2 1/2}$ harmonic quintessence quanta. By the same token if you want the mass of an object you need to multiply the quantum mass by that same number of quintessence quanta. Without further ado lets go on to derive the mass of the electron – for as we shall find, this is perhaps the most important mass unit in the Universe.

Box 4
<u>Mass of the Electron m_e in kg.</u>

taking c as the number (n) of harmonic quanta/sec

$^{\dagger}m_e = m_q\,(c^{2 1/2}/4\pi\,*)\;=\;9.109 \times 10^{-31}$ kg.

actual $m_e\;=\;9.109 \times 10^{-31}$ kg.

---

† Dimensionally m_q = [M.T] multiply by n = $c^{2 1/2}$, which is the number of light quanta *per unit time*, with the dimensions of [T⁻¹] we get the dimensions [M.T] × [T⁻¹] = [M].

> where m_q is the quintessential mass quantum = h/c^2, c is the number of speed of light quanta per second, ε denotes the standard electron magnetic moment to Bohr magneton ratio and * = (1 + 2ε).

Q.E.D. The mass of the electron - from the speed of light.

So the mass of the electron arises form first principles, from the light speed harmonic. Not only do we get the mass of the electron, but we get an explanation of how the Compton wavelength is derived. We also get an good idea of what is known as the anomalous magnetic moment of the electron. Experiment has shown the electron's actual magnetism (gyro magnetic moment) differs from the theoretical by a small but significant amount. In this model, it turns out that this can be explained by the fundamental structure of the electron. Now that we see the electron is made up of a number of quintessential harmonic quanta – it should have what is known as a binding factor. That is the amount of quintessential quanta that are given up when the electron came to form. So we get the explanation for three things about the electron from one precept, the wavelength, the mass, and the anomalous magnetic moment. Moreover in a later chapter (see: The Music of the Spheres) we can also calculate the charge of the electron and the fine structure constant from the very same ideas.

Moreover, what we will find is that the electron is the basic fundamental particle of nature and we can build the other particles, such as the proton, from it. We can liken this to the formation of currency, so if you take the basic mass quantum, this is like the cent, then if you take 100 hundred of these you get the dollar, which acts as you next unit of currency. It is very similarly with the electron, except there are about 10^{20} or 100,000,000,000,000,000,000 (actually $c^{2½}/4\pi$) quanta in our next unit of currency, the electron. So from the square root of a sphere with the radius of the inverse of the speed of light we can get the mass of the electron. We just need to multiply the quintessential quantum by the equivalent factor of the speed of light. The fascinating thing is using the same principle we can also define the charge of the electron from first principles and other mysterious constants of nature, (See Chapter 5, The Music of the Spheres) from the same principles.

Once we start down this road then virtually all the physical constants of nature may appear form these same first principles in a magnificently elegant way. What has been and shall be further demonstrated, in the words of the eminent physicist John Wheeler, *"is so compelling so beautiful that we will say to one another, yes, how could it have been any different."*

Chapter 5

The Music of the Spheres

There is geometry in the humming of the string.
There is music in the spacing of the spheres.

Pythagoras (569-475 B.C.)

Pythagoras was a great lover of music. Music itself is inherent in so many things in Nature. Not only is it present on the level of beauty that human music can achieve, but it is appears throughout the structure of the Cosmos. Music itself may not be present simply in the form of sound, but in the form of rhythm there is a rhythm in life itself as there is in many things in the Universe. There is rhythm in the beating of a heart. There is rhythm and motion in the waves of the sea and the tides of the oceans. There is rhythm in the heavens, this is manifest in the turning of the Earth on its axis, to the turning of the Earth around the sun, to the turning of our solar system within the spiral arms of the galaxy. The galaxy itself is home to a billion such solar systems all turning with a rhythm, which is part of Nature. Where does the underlying reason for this inherent rhythm come from? It comes from the very fabric from which the entire physical Universe is constructed -the quintessential oscillating quantum.

It was Pythagoras who saw the celestial bodies as following some musical pattern, with the stars and planets imprinted on crystal spheres, which hummed at a certain frequency. In truth the spacing of the celestial planets is not entirely dependant on some form of regular pattern, but based on their mass and velocity around the sun. Yes the spacing increases at every stage but it is also based on the density of the planets and chance collisions within our solar system.

However if you go down to a small enough scale then the mass spacing between the fundamental subatomic particles of the Universe is based on harmonics, and indeed we reintroduce the music of the spheres, but on the smallest possible scale; that of the subatomic particles and the quintessential ephemeral quanta from which they are themselves constituted.

One may commence the process of identifying the fundamental constants of nature from the understanding we elaborated in Chapter 3. The most important thing is that the constants are not ad hoc they are a logical consequence of understanding science. Once you understand where one constant comes from you can work out the rest. This is done is by working on the basis of the quintessential quantum of the universe. Not only is the velocity of this ephemeral quantum directly based on the speed of light, but in Chapter 3 it was shown that the wavelength was also directly related to the speed of light. It turns out that the speed of light is far more important than just the speed of light. With this association we can begin to open a window on the Universe, which gives an unimaginable insight into its workings.

One of the other most important constants of nature is the charge of the electron, if we had a handle on why the charge of the electron is what it is, the beauty of the elegance of the Universal design would begin to be revealed. The interesting thing about electrical charge is that there is essentially only one fundamental charge of the Universe. The electron has that charge and the proton has an equal and opposite charge.

The true hidden nature of the electron may recently have been revealed. Up till very recently it was considered that you could not divide the charge of an electron. However, new experiments show if you "squeeze" an electron hard enough (technically you need to do this with a magnetic field) then you can get a (quantum Hall) effect where you produce "quasi" electrons whose charge is 1/3e. Three scientists, Horst Stormer, Robert Laughlin and Daniel Tsui, who recently won the Nobel Prize for their work, have described this effect. They describe this quasi electron as a particle bound in one dimension, lets say the x direction, but not bound in the other directions, y and z, allowing dispersion in space as a vortex – or a sort of flattened whirl pool effect.

If we take three of these together, this would by all accounts form a three dimensional vortex, which would give the electron its charge. This is precisely what matter is fundamentally made of. In whatever way it is described, then the electron seems to be composed of three of these quasi electrons. This is a very important observation as it gives a clue as to the basis of the charge of the electron. Let us imagine under normal conditions, the three quasi electrons together as forming a sphere, which is the obvious structure for a particle. The equation of the charge of the electron would then depend on the equation for the volume of a sphere multiplied by 3, [mathematically,

$3V = 3(4/3\pi r^3) = 4\pi r^3$]. Now given that the quintessential quantum of the Universe has the velocity of the speed of light and a wavelength, which is also related to the speed of light, then one might expect the fundamental charge to be also related to this speed of light. Indeed it may turn out to be related to the volume of a sphere with the radius of the speed of light, in the most exquisitely elegant way (see Box 5)

Box 5
Fundamental Charge (e)[‡]

$$e = * \left[\frac{\varepsilon_0}{3(4/3\pi c^3)} \right]^{1/2} = 1.60218 \times 10^{-19} \text{ C}$$

or

$$e = * \left[\frac{\varepsilon_0}{4\pi c^3} \right]^{1/2} = 1.60218 \times 10^{-19} \text{ C}$$

actual e = 1.60218×10^{-19} C

where ε_0 is the electric constant, c is the speed of light, ε is the standard electron magnetic moment to Bohr magneton ratio, $* = 1/(1+2\varepsilon)^4$ and C is the unit of charge in coulombs.

Moreover we achieve considerable accuracy, as the actual charge of the electron to five decimal places gives exactly the same result (1.60218×10^{19} C). Mathematically we have something remarkable, the

---

[†] In dimensions * can be taken as the spherical binding energy with the dimensions of ML^6T^{-5}.

term $4/3\pi r^3$ is none other the volume of a sphere. So the charge of the electron is none other than the electric constant divided by three spheres each with a volume based on the speed of light. This is truly the music of the spheres, but this time the music is dependant on light, these are musical light spheres.

But wait, I hear some of my learned readers say, what about the dimensions[†]. This is part of the crucial paradigm shift. Not only is space and time united to from a unified whole space-time as with Einstein's physics, but energy-space-time are unified, and thus in this case, so is charge-space-time unified. This topic will be elaborated upon in a later chapter, suffice it to say this approach is immensely powerful. The great physicist John Wheeler, was convinced that a comprehensive theory of physics would depend ultimately on geometry and indeed he was correct.

So lets go on to the next stage of understanding this compelling beauty. There is another constant which effectively determines the speed of the electron round an atomic nucleus. The constant, called the fine structure constant, may also be derived based on these musical light spheres, in the same exquisite fashion. To date nobody has really understood where the fine structure constant comes from, there is a complex equation for its derivation based on other constants, but all they really know is that it is very important as far as the electron is concerned. In actual fact the fine structure constant is again based on the quintessential mass described here (see Chapter 3) and on the geometric structure of the electron, based upon the

---

[†] In dimensions * can be taken as the spherical binding energy with the dimensions of ML^6T^{-5}

concept of the light sphere (see Box 6). Specifically it is based on a sphere with a radius of the inverse of the speed of light.

Box 6
The Fine Structure constant (α)[‡]

$$\alpha = \frac{2\pi *}{m_q[3(4/3\ \pi c^3)*]^2} = 7.29739 \times 10^{-3}$$

or

$$\alpha = \frac{2\pi *}{m_q[(4\pi c^3)]^2} = 7.29739 \times 10^{-3}$$

actual $\alpha = 7.29735 \times 10^{-3}$

where $m_q = h/c^2$, ε is the standard electron magnetic moment to Bohr magneton ratio, $* = 1/(1+2\varepsilon)^8$, and c is the speed of light,

---

This is very close to the actual value of the fine structure constant 7.29735×10^{-3}. Yet again, I hear some of my learned readers say, what about the dimensions[†]. Once more this is part of the crucial paradigm shift, energy-space-time are unified. Suffice it to say that fine structure constant is formulated on basis of the musical sphere and the harmonics of the speed of light, and shows the same beautiful and compelling mathematical structure that of the charge of the

---

[†] In dimensions * can be taken as the spherical binding energy with the dimensions of $ML^6 T^{-5}$.

electron does. So here we have an explanation for the charge of the electron and the fine structure constant, these are again based on the exquisite geometry and symmetry of the Universe.

Chapter 6

The Spirit of the Atom

Without music life would be a mistake.

Friedrich Nietzsche (1844-1900)

Many great scientists and philosophers were enamoured by the beauty of music. Plato and Pythagoras played the lyre, Einstein the violin. It is as though there is a link between music and the mathematics of the Universe. String theory is based on the music generated by a vibrating string. Yet string theory has not yet developed the capacity to predict what occurs in nature. So far the new quintessence has predicted, the wavelength, mass and magnetism of the electron. It has also predicted the charge of the electron a very important constant, what is known as the fine structure constant, all from the speed of light.

We can do the same with the proton and begin to understand its (Compton) wavelength, mass and anomalous magnetic moment, by building it from the electron. Having established that the number of quanta present in the electron relates to the frequency of the electron, one just needs to find the right light speed harmonic for the proton, and the answer is that it is actually based on the square root of the speed of light ($c^{1/2}$). So we can find the harmonic ratio between the Compton wavelength of the electron and the Compton wavelength of the proton and in turn the ratio mass of the electron and the proton. Now if we go back to Chapter 2, you may recall the introduction of the term quark. According to quark theory, three quarks are required to make up the proton. So taking this into account then we are not surprised then that the actual harmonic ratio of the proton to the electron is related to the square root of the speed of light divided by 3 ($c^{½}/3\pi$). Because we are comparing the mass of the electron to the mass of the proton in this case the proton mass is the mass of the electron directly multiplied by this light speed harmonic.

Now in physics there are a specific set of quantities known as dimensions. For instance the dimension of mass is designated as [M], that of length [L] and that of time [T]. Now in any equation these have to balance. The beauty of these equations is the they do match in the following equation for the derivation of the mass of the proton. {For those who have a specific interest in dimensions, it is to be stressed that $c^{½}/3\pi$ *is the number of quanta per unit time. It does not* have the dimensions of the square root of the speed of light, but as before, it is the number of quanta per unit time it therefore has the dimensions of frequency [T^{-1}]. As we have already multiplied the quintessential mass by this time factor, when deriving the electron previously (Chapter 4), we need not repeat this from a dimensional standpoint and the light speed harmonic ($c^{½}/3\pi$) is treated as a pure number}.

Moreover, this light speed harmonic gives us surprisingly accurate values for what is termed the Compton wavelength and mass of the proton (see Box 7).

> **Box 7**
> **Proton Compton Wavelength ($\lambda_{C,p}/2\pi$)†**
>
> $\lambda_{C,p}/2\pi = (\lambda_{C,e}/2\pi) \div c^{1/2}/3\pi^* = 0.210308 \times 10^{-15}$ m
>
> actual $\lambda_{C,p}/2\pi = 0.210308 \times 10^{-15}$ m
>
> **Mass of the proton (m_p)† in kg.**
>
> $m_p = m_e \times c^{1/2}/3\pi^* = 1.672621 \times 10^{-27}$ kg.
>
> actual $m_p = 1.672621 \times 10^{-27}$
>
> $\lambda_{C,e}/2\pi$ is the Compton wavelength of the electron, c is the number (n) of speed of light quanta, ε is the standard proton magnetic moment to Bohr magneton ratio, $^* = (1 + \pi\varepsilon/9)$ and m_e is the mass of the electron.

Q.E.D. The Compton wavelength and the mass of the proton - from the speed of light.

So here we have it, the derivation of the two most fundamental particles of the Universe, the mass of the electron (Chapter 4) and the proton derived here, from first principles, from the concept of a light speed harmonic. We also get the fundamental charge and what is known as the fine structure constant (see Chapter 5). At the same time we also get the Compton wavelength of these particles. Additionally we get a

---

†Dimensionally, as we have already multiplied the quintessential mass by the number of quanta per unit time [M.T] x [T⁻¹] = [M], when deriving the electron, we need not repeat this from a dimensional standpoint and the term $c^{1/2}/3\pi$ is treated as a pure number.

clearer picture of the anomalous magnetism (proton magnetic moment to Bohr magneton ratio) of these particles, specifically this can be explained by the number of quanta that are lost upon the formation of the proton from these quintessence quanta. In the case of the proton and the other subatomic particles in this group, this depends on the fact that the proton is itself made up of three particles, known as quarks. Such that the actual ratio of each of these quark particles to the mass of the electron depends on a light harmonic divided by 3 [mathematically, $(c^{1/2}/3\pi) \div 3 = (c^{1/2}/9\pi)$]. So now we *can* explain the *effective* mass of each quark.

So why is number of the square root of the speed of light $(c^{1/2})$, so important in this case. Well if we briefly go back to what is known as the Schrödinger wave equation (which was incidentally derived from first principles in Book 1) we get the answer. Interestingly the term $(c^{1/2}/\pi)$ is a very special solution to the Schrödinger wave equation for the electron. Interestingly this term is the inverse of a standard solution to the Schrödinger wave equation for an electron confined in a space with radius of the speed of light (See box 8).

Overall, so much is clear, the precept used here is concise, elegant and logical, and is in agreement with the standard equations, which govern the motion of the electron. First we multiply the basic quantum of the Universe, the quintessence mass, by a specific light speed harmonic $(c^{2½}/4\pi^*)$ to get the mass, and Compton wavelength of the electron. Then the mass of the electron is multiplied by another special light speed harmonic $(c^{1/2}/3\pi^*)$ to get the mass of the proton. The proton is then made up of a total number of light speed quanta of ($c^3/12\pi^2$ *). This is then the total number of quanta contained in the proton per unit time and if we multiply this by the fundamental quintessence mass (see Chapter 3), we get the mass of the proton.†

Box 8
Derivation of the electron to proton mass ratio from Schrödinger

The standard equation for an electron confined in a one dimensional box is given by:

$$E\psi(x) = -\frac{\hbar^2}{2m} \cdot \frac{d^2\psi(x)}{dx^2}$$

If the one dimensional box has a length 2L, the quantum amplitude (A) can only be non zero between $x = 0$ and $x = 2L$ and the solution for the amplitude is:

$$A = (1/L^{1/2})$$

$$\psi(x) = (1/L^{1/2}) \sin \pi x/L$$

Where $L = c$ and $\sin \pi x/L = 1$ and multiplying by π

$$\psi(x) = (\pi/c^{1/2})$$

For three dimensions

$$\psi(x) = (3\pi/c^{1/2})$$

---

† Dimensionally m_q = [M.T] multiply by n, which is the number of light quanta *per unit time*, with the dimensions of [T⁻¹] we get the dimensions [M.T] × [T⁻¹] = [M].

These and many more constants may be very accurately derived from these very ideas. I hear some purists note that the mass of the proton is effectively only derived to 6 decimal places. Well all that needs to be said to these purists is that this new model is 1,000 times more accurate than the mass of the proton predicted by what is known as the current Standard Model (see Chapter 7). Moreover, we may accurately and rather magically go on to derive the effective masses of all the subatomic particles from these ideas.

Chapter 7

The Standard Model -Remodelled

All these constructions and the laws connecting them can be arrived at by the principle of looking for the mathematically simplest concepts and the link between them.

Albert Einstein.

In modern physics there a number of areas where the increasing complexity of the constructions and the laws connecting them are ever growing. What is required is the mathematically simplest concept to connect them. In the previous book the concept of a minimum quantum of mass, which matches in size to the minimal quantum of energy, Planck's constant, was introduced. Add to that the concept of mass being related to the harmonics of the speed of light, we can take physics far further than the standard textbook.[14] Firstly it is important to show the simplicity and accuracy of the new concepts, compared to the Standard Model. Whilst the Standard Model has been enormously successful, in truth we are faced in places in the physics textbook with some theoretical approximations. (See Box 9 and 10). In the standard textbook, for the mass of the up quark in the formation of the proton for instance the *effective* theoretical value is given as 362 MeV[†], where the effective mass used in the actual calculation of the proton mass is 336 MeV. Of these approximate

---

[†] MeV is another term used to describe mass and refers to mass in standard electron volts, strictly it is MeV/c^2, as you would expect from the quantum mass model.

theoretical values (362 MeV), we are told, "*these are rather close to the constituent quark mass, [336 MeV], derived from the magnetic moments.*"

Box 9
Standard Model Theoretical Constituent Mass of the Up Quark Mev/c²

For the Baryons: $m_u \approx 1/3(m_p + m_\Delta/2) = 362\ MeV$

For the Mesons: $m_u \approx 1/2(m_\rho + m_\pi/4) = 307\ Mev$

Actual: Using magnetic moments:

$$m_u \approx m_p/2.79 = 336\ MeV$$

When we do use the effective quark mass derived form the experimental "magnetic moments" to calculate them, as in the Standard Model, these are also a little off the mark. A few examples are given in Box 10. The mass of the proton for instance looks fairly accurate, but the mass of the other proton like particles (known as baryons) have long and complicated formulae with extra parameters, which are difficult to explain. A little disappointingly when one does the calculation for the mass of the proton, from the parameters given we get a mass equivalent of 940MeV, not 938 MeV, maybe not a big difference, but also not the twelve decimal place accuracy to which physicists refer to when describing the achievements of standard model. There is no doubt of the genius which inspired the model and the great successes it has enjoyed. Yet for some physicists the quark model is a phenomenological model, one that explained the findings but did not explain what lay underneath them (see Box 10).

Box 10
Standard Model Theoretical Baryon Masses in MeV/c²

Baryon	Formula	Predicted	Actual
$p = 3m_u + \delta' - \delta^3/m_u^2$		940	938
$\Lambda = 2m_u + m_s + \delta' - \delta^3/m_u^2$		1111	1115
$\Xi^- = m_u + 2m_s + \delta' - (\delta^3/m_u^2)(4r/3 - r^2/3)$		1323	1321
$\Sigma = 2m_u + m_s + \delta' - (\delta^3/m_u^2)(4r/3 - 1/3)$		1178	1197
$\Delta = 3m_u + \delta' + \delta^3/m_u^2$		1233	1232
$\Sigma^* = 2m_u + m_s + \delta' + (\delta^3/m_u^2)(1/3 - 2r/3)$		1372	1384
$\Xi^* = m_u + 2m_s + \delta' + (\delta^3/m_u^2)(4r/3 - r^2/3)$		1517	1529
$\Omega^- = 3m_u + \delta' + \delta^3/m_u^2$		1668	1672

Parameters used: $m_u = 336$, $m_s = 509$, $r = m_u/m_s = 0.66$, $\delta' = 79$, $\delta = 255$ MeV.

The other problem with the standard model is that one has to introduce fractions of charges of the electron, such that the up quark actually has a fractional charge of 2/3e and the down quark has the charge of -1/3e. Moreover all the quarks are restricted to this pattern. Thus in the standard model the proton with a charge of

1e is made up of 2 up quarks and a down quark so the final charge is +1e; [uud =2/3e + 2/3e − 1/3e = +1e].

So is it possible to give a more accurate account of the Standard model? Well, the mass of the proton has already been calculated to a far higher degree of accuracy than the standard model (see Chapter 6). So the answer is yes, but we have to escape the textbook *"phenomenologically motivated model"* and use a model, which is derived from first principles. The beauty of this new model is that it is based on a single fundamental particle of the physical Universe, the quantum that we described in Book 1, the new quintessence. This is then directly multiplied by a value dependant on the velocity of light and the geometry of the sphere. In this model this fundamental harmonic may give the mass of the electron from first principles (see Chapter 4). Then all we do is multiply the mass of the electron by a harmonic number, again based on the speed of light and we get the mass of the proton and other particle masses and so on and so forth.

Moreover, in particle physics the proton is far more important than the other particles in its group, so it is very important to get the mass of the proton correct. Here the mass of the proton will be calculated (in Mev/c²) to a high degree of accuracy. Form these calculations it is possible to show that the proton arises from very direct principles, initially based on the mass of the electron. So the number of quanta in the proton also relates to the speed of light (see Box 11).

Box 11
Standard Model -Revised Baryon Masses in MeV/c²

Baryon Formula[†]	Predicted	Actual
$p^+ = m_e \, (c^{1/2}/{*}3\pi)$	938.272	938.272
$\Lambda = \Sigma^- \div \pi^{1/16}$	1115	1115
$\Xi^- = \Delta \times \pi^{1/16}$	1323	1321
$\Sigma^- = \Lambda \times \pi^{1/16}$	1197	1197
$\Delta^+ = m_e \cdot (c^{1/2}/\,'3\pi^{3/4})$	1232	1232
$\Sigma^* = \Sigma^- \times \pi^{1/8}$	1381	1384
$\Xi^* = \Xi^- \times \pi^{1/8}$	1526	1529
$\Omega^- = m_e \, (c^{1/2}/{`*}3\pi^{1/2})$	1672	1672

c is the number (n) of quanta, ε is the standard proton magnetic moment to Bohr magneton ratio, $* = (1 + \pi\varepsilon/9)$, $' = (1 + 3\pi^3\varepsilon/9)$ $`* = 1/(1+\pi^3\varepsilon/9)$ and m_e is the mass of the electron in MeV

Q.E.D. The mass of the baryons - from the speed of light.

[†] Also $\Omega^- \div \pi^{1/12} = \Xi^* \div \pi^{1/12} = \Sigma^*$, and $\Xi^- \div \pi^{1/12} = \Sigma^-$

Dimensionally, as we have already multiplied the quintessential mass by the number of quanta per unit time, [M.T] x [T⁻¹] = [M], when deriving mass of the electron, we need not repeat this from a dimensional standpoint and the term $c^{1/2}/3\pi$ is treated as a pure number.

Here we have not only accurately derived the mass of the proton but also the other particles in the series called baryons (See Box 11). So we get orders of magnitude greater accuracy than the Standard Model and at the same time, all the approximations and many delta factors and the r factors, which appear in the standard model (see Box 9 and 10), are now unnecessary. So how do we get this degree of accuracy, well the Standard Model needs to be remodelled in a way that makes it more logical. Now, it is possible to operate on the basis of an effective mass of each quark in the proton as directly being one third that of mass of the proton. This approach has a certain beauty, as each quark in the proton is now on an equal footing. So the proton now is made up of three up quarks (uuu) each with the charge of +1/3e of the standard charge. So we no longer need the messy 2/3and -1/3 pattern.

So what is the proton? Well you may recall in Chapter 5 we introduced the concept of the -1/3 fractional negative charge of the electron, which in physics is termed a quasi electron. The mirror particle of this is called the quasi positron with a charge of +1/3 fractional positive charge. So each quark is a tri-quasi positron complex, multiplied by a square root light speed harmonic, which is importantly is in keeping with the standard equations of quantum electro-dynamics (Q.E.D. see Box 8)

How then do we derive the charges of the other subatomic particles? For instance in the Standard Model the neutron with no net charge is defined as (udd); so doing the calculation shows that the charge is zero (2/3e-1/3e-1/3e = 0). In the new model the neutron is a complex of the proton with the total charge of +1e, effectively tightly orbited by an electron of charge −1e, which makes the net charge zero (see Box 21, Chapter 12). So how is this possible, when we are taught that in such circumstances the electron would effectively disintegrate, due to the massive orbital accelerations it would be under in such a confined space. This in reality is the same question as why does the electron orbiting an atomic nucleus, as in a hydrogen atom, not disintegrate? The standard answer is that the electron orbitals are quantized. However, as with many aspects of quantum physics this is just a reason not to think about the problem, nor to question it.

Well the answer to both is straightforward once you understand the *nature of charge*. Recently scientists are beginning to realise that charge is a vortex. To be exact it is a vortex of space-time. Now imagine that you happen to be sitting in such a vortex, it is the space-time around you that is being accelerated so in actual fact you do not feel the acceleration, although you will no doubt be going at some speed. The same applies to the electron in close orbit around a proton to form a neutron. Having said that, the neutron is not a very stable particle and outside an atom it does disintegrate in about 15 minutes principally into a proton and an electron.

This begs the question, why is the neutron so stable when it is *inside* the atom, or even why are atoms so stable? After all, how can a neutral particle such as the neutron hold all those strongly repelling positive charges together? The answer is now direct, the neutron is positive on the inside and effectively negative on the outside (see Fig 2, Chapter12), and in this way it practically glues the protons in the atom together. Indeed recent experiments confirm this view of the neutron as been more positive in the middle and more negative on the outside.

The next question must be, why is the mass of the neutron what it is? Well in truth the standard model has no accurate answer to this question, except that the down quark weighs a little more than the up quark. In actual fact the neutron actually weighs in at a little more than the proton (at 939.565 MeV). This is equivalent to the mass of the proton plus about two and a half times the mass of an electron. This is where relativity enters the particle puzzle, because the central proton is about 2,000 times heavier than the electron the electron effectively orbits the proton very closely and thus at great speed. If we calculate this speed it is velocity is approximately 91.863% the speed of light (see Box 21) and at this velocity due to well-known relativistic effects its mass increases to about 2.53 times the mass of the electron as observed. Exactly what you would expect for the experimental mass of the neutron, derived using relativity.[16,17]

Following the same pattern in this way we can also accurately calculate the exact masses differences of the baryons (that is particles made of three quarks) with their differing charges. So in the baryon particle sequence the particle with the positive charge is the lightest, the neutral particle is slightly heavier, with one additional electron, and the negative particle is slightly heavier still, with a pair of additional tightly orbiting electrons. This slightly different than in the Standard model, but once one understand the nature of charge it makes absolute sense. Importantly it also explains the decay pattern for instance of a neutron which effectively decays in to proton and electron. (see Chapter 12, Fig 2 and Box21).

We can elaborate upon this for clarity. Lets start with a question, which the standard model does not seem to be able to explain adequately. Why do some particles with three quarks (e.g. Σ^-) end up with two decay products that contain a total of five quarks (e.g. $\Sigma^- \rightarrow \Lambda$ and π^-)? Well the answer here is straight forward the negative particle already contains 3 quarks and 2 proto quarks in the form of a pair of closely orbiting electrons. So this particle readily decays in a tri-quark (baryon) particle and a di-quark (meson) particle. And so the pattern continues with the negatively charged baryons. The next particle in the sequence, (Xi⁻), does exactly the same ($Xi^- \rightarrow \Lambda$ and π^-). The next baryon in the sequence also decays into five quarks, but this particle is heavier and this time the di-quark particle can afford to be a bit heavier too ($\Omega^- \rightarrow \Lambda$ and K^-). So we get a good explanation for the decay products using this model. (See also Chapter 12). But the important thing to note is that theses particles ultimately all decay into an electron (or its mirror partner the positron). In music terms the electron is the *fundamental* harmonic.

Similar to the standard model the delta particle (Δ) contains three down quarks (ddd) each with the one-third charge of the electron. The omega (Ω) particle contains three strange quarks (sss). But in effect these quarks and the other particles are really just pi (π) harmonics of the proton (see Box 11). Now in quantum physics no two identical particles are allowed to coexist together. This initially caused some difficulties, but actually there is a further characteristic which physicists discovered, which was called colour after the three primary colours of light, red, blue and green. Thus the description of these particles is often called Quantum Chromo Dynamics (QCD). In fact it is not only colour, which determines these differing characteristics it is the perpendicular vector, after all there are three dimensions, where the quark occupies either x, y or z vector, and it should therefore be called Quantum Vector Chromo Dynamics (QVCD).

There are other particles called mesons, which are basically quark doublets, specifically two quarks together, the masses can be calculated much the same as in the standard model. The calculations were already logically dependant on a quantity known as m_0, where that quantity given (m_0 = 311 MeV) is pretty much the value of an up quark. That is one third the value of the proton ($m_p/3 = c^{½}/9\pi^* $ =312.757 MeV) In this case I think it would be reasonable to say *"these are rather close to the constituent quark mass"*. So this part of the model now makes more sense as the important value, the mass of the proton (m_p), is more accurate and we do not need those other two delta and r factors used in the standard model of the baryons (see Box 10 and 11).

Thus, the actual ratio of each of the mass of these individual constituent quark particles to the mass of the electron (actually the mirror image of the electron called the positron) depends on a light speed harmonic. Each of the constituents quarks in the meson model is now based on the effective mass of the up quark. Using this mass, we can arrive at the correct meson masses. (See Box 12). This system is more accurate that the standard model and by taking into account the proton magnetic anomaly we also get a slightly more accurate value for the effective mass of the quark related to its magnetic moment (m_u = 337).

What we find with these light mesons is that a lot of the fundamental quark mass disappears, as the particles orbit each other and the remaining mass is determined by what is known as the magnetic moment. This is like two dancers holding each other at arms length and spinning, they are effectively supporting each others weight. This is recognized by conventional physics and is called spin splitting, which is part of the standard model. By re-introducing this concept into the quark model above we get an accurate model of the particles known as mesons in a logical and also conventional way. The strength of this new approach is that it uses this same information for calculating the baryon masses as it does for calculating the meson masses and shows how fundamental the proton mass is to calculating both these sets of particles in a more accurate way than the standard model (see Box 12).

The real problem with the old quark model is that it did not explain why the masses of the quarks are what they are. Nor did it explain where the electron mass came from. Indeed, we can now explain where both of these come from on the same basis. If we take the mass of the electron (itself based on a spherical harmonic of the speed of light) and multiply it by another light speed harmonic, which is in keeping with the equations for quantum electrodynamics, (Q.E.D. see box 8), we get the mass of the proton and if we take one third of this mass we get the effective mass of the individual quarks. Using this mass, we can also readily calculate the masses of all the particle made of two quarks, called mesons on this basis, again from the same principles.

Box 12
Standard Model Revised- Meson Masses in MeV/c²

Meson	Formula	Predicted	Actual
$\pi =$	$2m_u - \dfrac{2(m_p/3)^3}{m_u^2}$	135	135
$K =$	$m_u + m_s - \dfrac{2(m_p/3)^3}{m_u \, m_s}$	494	494
$m_{ss} =$	$2m_s - \dfrac{2(m_p/3)^3}{m_s^2}$	790	~790
$\rho =$	$2m_u + \dfrac{2(m_p/3)^3}{3m_u^2} - \Delta$	771	771
$K^* =$	$m_u + m_s + \dfrac{2(m_p/3)^3}{3m_u \, m_s} - \Delta$	885	892
$\phi =$	$2m_s - \dfrac{2(m_p/3)^3}{3m_s^2} - \Delta$	1019	1019

where $m_p/3 = 312.757$, is the mass of the proton/3, m_u is given as 337 and m_s is given as 512 and the standard term $\Delta = 82$

Q.E.D. *The meson masses*

The standard model calculations, which express why the mass of the proton is what it is, is only accurate to about two decimal places. The new model allows its calculation to five decimal places, particularly in the case of the proton and neutron (see box 21, Chapter12). Moreover, these particle masses are dependant on the masses of the quarks but nobody, to date, knew why the mass of the quarks are what they are. The problem had just been moved back one stage. The quark model *is* essentially almost correct but nobody realised why. The mystery was compounded by the resultant particles having an eightfold symmetry that is called the "eightfold way", after the Buddhist philosophy, which reveres the number 8. Now that symmetry drops out from the maths (see Box 11).

When a ship is sinking at sea, there is a tendency to repair it with whatever is to hand at the time. This means while the repair works, that is good enough. This is not to say that quark theory is not the product of the genius of the time, nor wrong, but the understanding behind it was missing. Quarks are formed from the harmonics of the speed of light multiplied by the mass of the electron, (which itself is based on a spherical harmonic of the speed of light). This might have been clear to us, because if you watch the breakdown of these particles they all eventually decay in to an electron (or its positively charged mirror partner the positron).

What is shown in this book is that the quark model can be understood at the most harmonious level. The link between quantum physics, relativity and particle physics is made and these three aspects of physics are beautifully wed. So much so, that the constants of nature and masses of the fundamental particles may drop out from first principles in the most elegant fashion, from a single fundamental quantum. The truth and beauty of the Universe then becomes quintessentially awesome.

Chapter 8

Vortex Harmonics

Entia non sunt multiplicanda praeter necessitatem – Entities should not be multiplied beyond necessity.

William of Ockham (1285-1349)

Occam's Razor as it has become know, is a very important tool of science, for it forbids the use of many explanations when one will do. Modern physics is desperately in search of a unifying theory of physics precisely because there are so many disparate fields in physics. What science needs is an overview, one explanation for the whole plethora of the laws of Physics.

In chapter 4 we outlined such an approach as being based on a single quintessential quanta. Taking the velocity of this quantum, as none other than the speed of light, it was possible to use this ephemeral mass quantum to derive the Compton wavelength and mass of the electron. This was done by showing that these parameters are based on a harmonic of the number of quanta, the harmonic was itself elegantly based on the speed of light. Using the mass of the electron as the "fundamental" harmonic of matter, then the mass of the proton dropped out from first principles also based on the speed of light. We similarly found the light harmonic for all the subatomic particles related to the proton. In so doing, we may have uncovered one of Nature's most precious secrets.

In the meantime we have also answered many of the secrets that Nature still held. Why is the charge of the electron what it is? What on Earth is charge anyway? Why is the fundamental charge of the electron equivalent to the charge of all the other charged subatomic particles. Specifically all charges are multiples of the charge of the electron. The answer to the first question is revealed in the Chapter entitled "The Music of the Spheres" (Chapter 6). The answer to the second question is charge is just like a vortex of space-time. Like any ordinary vortex, take two vortices spinning in opposite directions they will attract. Vortices spinning in the same direction will repel. The answer to the third question is that the electron is musically equivalent to what is known as the "fundamental" and all else are just harmonics of the fundamental. So we should not be at all surprised to find that there a number of specific "vortex" harmonics of the electron, but they all have the same charge.

The second expected vortex harmonic of the electron is that particle which we briefly mentioned before in chapter 2, is the muon. So surprised were conventional scientists to find that such a particle existed, that one eminent physicist of the time remarked:

"Who ordered that?"

Isidor Isaac Rabi.

The muon is a particle with the charge of the electron, but a higher mass. It is part of a group of particles called leptons, which are effectively electron like. In this group there are essentially three particles the electron itself, the muon and the tau particle. It will be shown here that these extra leptons are based on the third root otherwise known as the cube root of the speed of light. Now this is where it gets rather interesting because the charge of the electron, which itself results from the charge vortex (see Chapter 5, The Music of the Spheres), is in turn based on volume of a sphere with the radius of the speed of light cubed. So by simple analogy these particles all dovetail in with the charge vortex of the electron, precisely because they are dependant on the cube root. So we could say that these leptons are based on vortex harmonics. By this we means that the extra frequency generated by the extra numbers of quanta exactly match its charge.

Remembering also that there are three (tangible) dimensions it is not surprising that it is based on the third (cube) root. As the light speed velocity of the quintessential quantum suggests it should be, the mass and wavelength of this particle thus turns out to be based on the cube root of the speed of light. Specifically, the answer is that it is based on the cube root of the speed of light divided by π or mathematically $(c^{1/3}/\pi)$. As we are comparing the mass of the electron to the mass of the muon, in this case the mass is given by the number of light speed quanta multiplied by the mass of the electron. So using exactly the same principles we used before, accurately gives the Compton wavelength and mass of the muon (see Box 13).

Box 13
<u>Muon Compton wave length ($\lambda_{C\mu}/2\pi$)†</u>

$\lambda_{C\mu}/2\pi = \lambda_{C,e} /2\pi \div (c^{1/3}/\pi^*) = 1.8677 \times 10^{-15}$ m

actual $\lambda_{C\mu}/2\pi = 1.8676 \times 10^{-15}$ m

<u>Muon mass (m_μ)† in kg.</u>

$m_\mu = m_e . (c^{1/3}/\pi^*) = 1.8833 \times 10^{-28}$ kg

actual $m_\mu = 1.8835 \times 10^{-28}$ kg

where $\lambda_{C,e}/2\pi$ is the Compton wavelength of the electron c is the number (n) of light speed quanta, ε is the standard muon magnetic moment to Bohr magneton ratio, * = (1 + 2πε) and m_e is the mass of the electron.

Q.E.D. The Compton wavelength and the mass of the muon - from the speed of light.

Now everything is becoming increasingly clear we have the Compton wavelength and mass of the muon using the same entirely consistent first principles as were used to define the fundamental wavelengths and masses of the electron and proton (see Chapter 4). Additionally, we can also indirectly determine another important parameter (known as the muon magnetic

---

†Dimensionally, as we have already multiplied the quintessential mass by the number of quanta per unit time [M.T] x [T⁻¹] = [M], when deriving mass of the electron, we need not repeat this from a dimensional standpoint and the term $c^{1/3}/\pi$ is treated as a pure number.

moment to Bohr magneton ratio) and explain its existence on the basis of the binding energy of the muon, specifically that it is the energy that holds the muon together.

It would be indeed amazing if it were possible to explain the wavelength and mass of the third (tau) particle of this series on the basis of the same third root or cube root principle. Suffice it to say it is entirely possible and logical to do so. So here it is, we just need to take the third (cube) root of the speed of light and multiply it by the third (cube) root of the third (cube) root of the speed of light and the right answer pops up. Specifically, the equation for finding the mass of the tau depends the third root of the speed of light multiplied by the ninth root of the speed of light, both divided by π, or mathematically $[(c/\pi)^{1/3} \cdot (c/\pi)^{1/9}]$[†]. So we just multiply this figure by the mass of the electron, which we previously found, and *viola* we have the mass of the tau particle. Such absolute compelling elegance is not seen in the standard model of particle physics, which actually has no idea why the masses of these particles are what they are. So using the same techniques as before we get the right answer (see Box 14).

[†] Dimensionally, as we have already multiplied the quintessential mass by the number of quanta per unit time [M.T] x [T⁻¹] = [M], when deriving the mass of the electron, we need not repeat this from a dimensional standpoint and the term $(c/\pi)^{1/3} \cdot (c/\pi)^{1/9}]$ is treated as a pure number.

> **Box 14**
>
> <u>Tau Compton wave length ($\lambda_{C\tau}/2\pi$)†</u>
>
> $\lambda_{C\tau}/2\pi = \lambda_{C,e} / 2\pi \div *(c/\pi)^{1/3}.(c/\pi)^{1/9} = 0.111046 \times 10^{-15}$ m
>
> actual $\lambda_{C\tau}/2\pi = 0.111046 \times 10^{-15}$ m
>
> <u>Tau mass (m_τ)† in kg.</u>
>
> $m_\tau = m_e *(c/\pi)^{1/3}.(c/\pi)^{1/9} = 3.16777 \times 10^{-27}$ kg
>
> actual $m_\tau = 3.16777 \times 10^{-27}$ kg
>
> where $\lambda_{C,e}/2\pi$ is the Compton wavelength of the electron c is the number (n) of light speed quanta, ε is the predicted tau magnetic anomalous moment (0.001177929), $* = 1/(1 + \pi^3\varepsilon/3)$ and m_e is the mass of the electron.

Q.E.D. *The Compton wavelength and the mass of the tau - from the speed of light.*

So there it is, in a three dimensional Universe we have particles based on the third (cube) root of the speed of light, where the fundamental speed of the quintessential quantum is itself the speed of light† and

---

† Dimensionally, as we have already multiplied the quintessential mass by the number of quanta per unit time [M.T] x [T⁻¹] = [M], when deriving the mass of the electron, we need not repeat this from a dimensional standpoint and the term $(c/\pi)^{1/3}.(c/\pi)^{1/9}$ is treated as a pure number.

† Actually the product of the phase wave velocity x the actual velocity is c^2

the actual wavelength of an individual quantum is none other than the speed of light in meters.

What can be more compelling, that we can derive the fundamental masses of the particles by first multiplying the quintessential mass quantum by a harmonic of the speed of light to get the mass of the electron. By further multiplying the mass of the electron by a further harmonic of the speed of light it is possible get the mass of the other known particles. So what we have seen is a harmonious Universe, whose quintessential beauty is breathtaking.

Chapter 9

Higher Order Quarks

The grand aim of all science is to cover the greatest number of empirical facts by logical deduction from the smallest number of hypotheses or axioms.

Albert Einstein.

It was around 1970 and the particle mystery deepened. More and more subatomic particles were being discovered. These were heavier particles than previously found in the accelerators. But as the accelerators were getting more powerful, more particles of a heavier nature were being discovered. Where did all these heavier particle come from? To explain the new particles, three famous scientists named Glashow, Iliopoulous and Maiani, postulated there may be a fourth quark and named it *charm*. In fact it was George Zweig, who had originally postulated in 1964 that there were four quarks and called them aces. He should perhaps have been credited more at the time.

As it happens, four quarks were still not enough and we do now know that there are actually six quarks. These are required to explain the ever-bigger particles that have since been discovered. The problem with all this is if you count all the particles (some of which have not been mentioned as yet) and their mirror antiparticles and then if you add the forces that act between these particles, then in total you get about 72 fundamental particles and force mediating entities. So you still have a mysteriously complex picture of Nature

So far the work of this book has been to identify the single fundamental quantum of the Universe and base just about everything known about particle physics on this one quintessential mass. Moreover all this appears to depend on totally logical first principles. The next question is having found the lower mass particles, on the basis of this quantum mass; is it possible to do the same with the higher mass particles? We are of course constrained by the same single axiom. The answer is unless this axiom is the correct one, the chances of using it to find the higher quark masses are billions of billions to one against. Let's see if we can beat these odds.

In chapter 4, it was possible to show the derivation of the mass of the electron from the mass of the quintessential mass quantum, by multiplying it by a light speed harmonic. Then it was shown that the mass of the proton then dropped out from first principles by multiplying the mass of the electron using another light speed harmonic. In chapter 5, it was then possible to show that the mass of the proton and the lower order quarks and in turn the other lower mass particles all depended on these same light speed harmonics. In chapter 8 it was then shown that the higher order electron like particle (leptons) were based on vortex harmonics again based on the speed of light. So logically and elegantly the mass of the higher order quarks should be based on these higher order electron like particles. It turns out they are just that.

The next two quarks particles we find in the standard model are called charm and beauty (the latter sometimes sadly called bottom). The key to the masses of these two particles of the standard model is the mass of the next lepton up from the electron, specifically known as the muon (μ), which we describe in the previous chapter. Now the muon itself is an electron powered by a specific harmonic of the speed of light $(c^{1/3}/\pi)$. Not surprisingly when these quarks decay they decay in to exactly that a muon. So using this muon multiplied by a specific harmonic, like the one used to derive the proton from the electron, gives the right answer for the charm quark.

Box 15
Mass of the Charm Quark (m_c) in GeV[††]

$m_c = m_e (c^{1/3}/\pi)(c^{1/4}/3\pi) = 1.52$ GeV

Actual mass $m_c \sim 1.52$ GeV

Which is equivalent to

$m_c = \mu(^*c^{1/4}/3\pi)$

where c is the number (n) of quanta, m_e is the mass of the electron, ε is the standard muon magnetic moment to Bohr magneton ratio, * = (1 + 2πε) and μ is the mass of the muon.

[††]Dimensionally, as we have already multiplied the quintessential mass by the number of quanta per unit time [M.T] x [T⁻¹] = [M], when deriving mass of the electron, we need not repeat this from a dimensional standpoint and the term $(c^{1/3}/\pi)(c^{1/4}/3\pi)$ is treated as a pure number.

Q.E.D. The charm quark mass - from the speed of light.

So there you have it the mass of the charm quark from a light speed harmonic of the muon, which itself is a light speed harmonic of the mass of the electron, which is itself a light speed harmonic of the quintessential mass quantum.

You may well be asking yourself at this point why we don't have a more accurate value for the mass of the charm quark than this. Well, somewhat like very shy people, it is very rare for these quarks to be seen alone if at all. That is because the force binding them together is very strong and gets stronger the further apart you try to pull them. So we don't often see what is often termed in physics as bare quark masses and we have never, fortunately, seen a bare bottom quark. This means in effect that physicists have to measure their masses indirectly. In technical terms the masses are measured as RGI (renormalization group invariant) masses or quenched QCD mass. Both these are a very complex way of trying to measure mass.†

There is of course a far better way of measuring mass of the charm quark than this, that also potentially gives us a more accurate answer and explains where the mass comes from. Lets go back to the concept of mesons, which are two quarks together. The heavy mesons don't behave like the light mesons, precisely because their constituents are that of a muon multiplied by a harmonic of the speed of light. So the meson, made up of a charm quark and an anti-charm quark, will behave as if the anti-charm quark orbits around the charm quark. So the mass of the charm

---

† The above actual mass given in box 12 is an average of these values.

quark is easier to find, all we need to do is find this double charm particle and halve its mass (and take a little off for the relativistic effects and the binding energy of the charm itself). There is such a particle called the J/Psi, (J/ψ = 3.09 GeV) and it weighs in at just over twice the mass of the charm particle, pretty much as expected by this approach to getting the mass.

As regards the beauty quark, we can do almost exactly the same thing, by starting with the structure of the muon, and what we get is pretty much the same pattern (see Box 16). What the standard model has always failed to do is explain exactly why beauty and charm behave in similar ways by decaying in to muon and an anti-muon. In the standard model it is the strange quark and charm that occupy the same level, yet they behave very much differently. For our part we put the charm and beauty quarks on the second level along with the muon.

Box 16
Mass of the Beauty Quark (m_b) in GeV/c^2 [†]

$m_b = m_e \cdot (*c^{1/3}/\pi^{1/2})(c^{1/4}/3\pi^{1/2}) = 4.63$ GeV

$m_b = \mu(c^{1/4}/3) = 4.63$ GeV

Actual mass $m_b \sim 4.5 \pm 0.3$ GeV

[†] Dimensionally, as we have already multiplied the quintessential mass by the number of quanta per unit time [M.T] x [T⁻¹] = [M], when deriving mass of the electron, we need not repeat this from a dimensional standpoint and the term $(c^{1/3}/\pi)(c^{1/4}/3\pi)$ is treated as a pure number

> where c is the number (n) of quanta and μ is the mass of the muon.

Q.E.D. The mass of the beauty quark - from the speed of light.

You may again be asking yourself at this point why we don't have a more accurate value for the mass of the beauty quark than this. Well again the conventional measurements of the quark masses are indirect. It is however important to infer the direct masses from the direct measurement of known particles, so we can get the *effective* mass of the quark. Again there is a particle, which is made up of a beauty quark and an anti-beauty quark known as the Upsilon particle (γ = 9.46 GeV). In this meson the beauty quark sits in the middle and is orbited by the anti-beauty quark so the effective mass of the beauty quark is just under half this value (take a little off for the relativistic effects and the binding energy of beauty itself). As you might expect from its structure (See Box 16), the Upsilon particle decays into a muon and an anti-muon.

What then of the truth quark, well this belongs on the third (highest) level along with the tau lepton. If you look at the mathematical structure of the truth quark it is made from a tau particle again multiplied by a light speed harmonic. By following the previous pattern of particles, which we were also able to derive very neatly from starting at the quintessential mass, we can derive all the particles on this basis just by using the harmonics of the speed of light. Remarkably, by taking the mass of quintessence and multiplying that a light speed harmonic we get the mass of the electron, multiply that by another harmonic to get the muon, multiply that by similar harmonic and you get the tau particle and multiply that by another harmonic and you get to the truth.

Box 17
Mass of the Truth Quark (m_t) in GeV[‡]

$m_t = \tau (c^{1/4}/\pi) = 177.9\ GeV$

Actual mass $m_t \sim 178 \pm 4.3\ GeV$

Where c is the number (n) of light speed quanta, and τ is the mass of the tau lepton.

Q.E.D. The mass of the truth quark - from the speed of light.

So there you have it the mass of the truth quark from a light speed harmonic of the tau, which itself is a light speed harmonic of the mass of the electron, which is itself a spherical harmonic of the fundamental quintessential mass quantum.

Moreover if we watch the decay of the truth quark we can see exactly how the various particle inter-relate (see Box 23). Now we know why the masses of the particles are so distributed. In the new model both beauty and charm sit on the second level. The truth quark sits at the top of the tree of particles,

---

[†] [‡]Dimensionally, as we have already multiplied the quintessential mass by the number of quanta per unit time [M.T] x [T⁻¹] = [M], when deriving mass of the electron, we need not repeat this from a dimensional standpoint and the term $(c^{1/4}/3)$ is treated as a pure number.

on the third level, it is made from a tau particle multiplied by a light speed harmonic, thus its mass is so much greater than all the rest. It has the same mass as an entire gold atom made up of many, many protons and neutrons.

From this and the previous chapters we can now see where all the mysterious masses of the quarks and in turn all the other related particles come from. They can all be derived from a light speed harmonic multiplied the ephemeral quintessential mass. In the next Chapter, we lay all these particles out in such a symmetrical way that we can see at a glance the exquisite design of the physical Universe.

Chapter 10

It Must Be Aesthetically Beautiful

The most beautiful thing we can experience is the mysterious it is the source of all true art and all science.

Albert Einstein

In the early days of modern physics, what lay at the heart of the atom or the nucleus was in reality still unknown. The assumption was made that the neutron and proton where simply fundamental particles in their own right and not related to something far more fundamental than this. They had little knowledge at the time to complete the picture and part of that complete picture was of course, particle physics. Particle physics was only just being discovered at the time. The forces that were well known then, were gravity and electromagnetism (light), little was known about the mysterious force that holds the subatomic particles together, called the strong nuclear force.

We have in the previous chapters effectively been describing this strong force. Now for every force there is a force mediator. So powerful are the strong force mediators that they are termed gluons. These effectively glue the various particles like the proton together. In the model described here these force mediators or gluons all are special harmonics of the speed of light itself multiplied by the truly ephemeral quintessential quantum of nature. That is one of the things that makes this model so compelling. In mathematical terms this makes perfect sense. This is because if we take the velocity of the individual quintessential quantum which make up this force, then it has the speed of light and a wavelength related to the speed of light.

One of the most mysterious parts of physics is why the mass of the elementary particles are what they are. If we were to adhere to the standard model rigidly, we would probably never find out. However a revision of the standard model, as described here leads us directly to the answer.

We start with the three leptons (electron like particles). To recap the characteristics of the electron itself is based on the square root of a sphere with the radius of the inverse of the speed of light as is the charge of the electron (see Chapter 4 & 5). The next lepton in the series, the muon, is the electron multiplied by the cube root of the speed of light. The next lepton after that, the tau, is the electron multiplied by the cube root of the speed of light, multiplied again by the cube root of that cube root. What is so surprising is that once one sees the maths it is breathtakingly harmonious.

Again the accuracy of these measurement is stunning, provided we take in to account the magnetic moment, then we get five decimal place accuracy (see boxes 3, 4, 13 and 14 for the electron, muon and tau respectively). Not only do we get the mass we get the (Compton) wavelength of these particles and at the same time the binding energy explains pretty much precisely an anomaly in current physics (what is termed the anomalous magnetic moment). A summary of the lepton array is given in Box 18.

Box 18
Lepton Array[‡]

$$m_e = m_q (c^{2\frac{1}{2}}/4\pi^*)$$

$$m_\mu = m_e (c^{1/3}/\pi^*)$$

$$m_\tau = m_e *(c/\pi)^{1/3}.(c/\pi)^{1/9}$$

where c is the number of light speed quanta per unit time, m_q is the quintessential quantum mass, m_e m_μ and m_τ are the masses of the electron muon and tau respectively,. * is based on the particle magnetic moment to Bohr magneton ratio, see Boxes 3,4,13,14.

Q.E.D. The mass of the leptons form the speed of light

What we see here is the basic lepton array. The higher lepton masses essentially decay in to the lower lepton particle plus the release of a photon. So the muon will decay in to an electron with the release of a photon. The tau will decay into a muon, with the release of a photon exactly as one would expect from the lepton array. The light speed harmonic given for each of the higher leptons can thus be considered like a trapped photon, which has exactly the right (non rest) mass and thus frequency to dovetail in with the electron to give what is called in Chapter 8, vortex harmonics.

[‡] Dimensionally, as we have already multiplied the quintessential mass by the number of quanta per unit time [M.T] x [T⁻¹] = [M], when deriving mass of the electron, thereafter we need not repeat this from a dimensional standpoint and the factors of c are treated as pure numbers.

For our learned colleagues, another very tiny particle, which has not previously been mentioned, called the neutrino is also released from these decay processes. The thing is that the Standard Model has not got a clue what these particles are. Again this new model is so predictive that it can readily predict the structure and masses of these tiny neutrino particles, and it will be possible to elaborate on the nature of the neutrinos in the next chapter, using the lepton array.

Suffice it to say the lepton array can also be used to determine the quark array in what is a startling symmetry (see Box 16).

Box 19
Quark Array[‡]

$$u = m_e\,(c^{1/2}/{}^*9\pi) \qquad d = m_e\,(c^{1/2}/{}^*9\pi^{3/4}) \qquad s = m_e\,(c^{1/2}/{}^*9\pi^{1/2})$$

$$ch = m\mu\,({}^*c^{1/4}/3\pi) \qquad\qquad b = m\mu\,({}^*c^{1/4}/3)$$

$$t = m\tau\,(c^{1/4}/\pi)$$

where c is the number (n) of light speed quanta, m_e m_μ and m_τ are the masses of the electron muon and tau respectively, u, d, s, ch, b and t are the up, down, strange, charm, beauty and truth quarks, * is related to the particle magnetic moment to Bohr magneton ratio, see Boxes 11,15,16,17.

Q.E.D. *The quark array from the speed of light harmonics.*

[†] Dimensionally, as we have already multiplied the quintessential mass by the number of quanta per unit time [M.T] x [T⁻¹] = [M], when deriving mass of the electron, thereafter we need not repeat this from a dimensional standpoint and the factors of c are treated as pure numbers.

So the quark array now has a perfectly logical origin. To recap we start with the electron, which is itself derived from the fundamental quintessential mass quantum (m_q) multiplied by a spherical light speed harmonic (see Chapter 4). To get the mass of the proton we multiply the electron mass by a light speed harmonic, based on standard quantum physics ($c^{1/2}/3\pi$, see Chapter 6, box 8) [†]. To get the effective mass of the individual up quark we then divide this by a factor of three. These mathematical structures, multiplied by the mass of the electron represent the mass of gluons. Now to obtain the other quarks in this level, the down and strange quarks, we merely divide this up quark by $\pi^{1/4}$ and $\pi^{1/2}$ respectively, plus a tiny factor depending on the binding energy of the quark particle (see Box 18). There are two more gluons at this level, to get these and the rest of the particles made up from the quarks at this level, we merely divide by $\pi^{1/8}$ and $\pi^{1/16}$ (see box 11).

The next two quark particles, charm and beauty, are on the second particle level and are represented by the muon multiplied by a further two gluons, these have the gluon mass with the square root of the square root of the speed of light ($c^{1/4}/3\pi$) and to get the second gluon on this level we merely effectively divide by π (actually the muon component *and* the gluon component are each divided by $\pi^{1/2}$ in keeping with the first level of particles see Box 16). So if we multiply these numbers by the mass of the electron we get the

---

[†]Dimensionally, as we have already multiplied the quintessential mass by the number of quanta per unit time [M.T] x [T^{-1}] = [M], when deriving mass of the electron, thereafter we need not repeat this from a dimensional standpoint and the factors of c are treated as pure numbers

effective mass of these gluons. How can we tell that this is the structure, well both these quarks decay giving the same decay product, and that is the muon from which they arise.

The next quark, the truth, resides on the third level and is derived from a tau particle complexed with the last gluon ($c^{1/4}/\pi$). How can we tell this is the structure, well if you watch the decay of the truth quark one of the pathways for its decay is via the beauty quark (see Box 23, Chapter 12).

This gives us the total of eight gluons we know that exist in experimental physics. So with these eight gluons we can account for all the subatomic structures of this class of particle, the tri-quark particles (known as the baryons) and the di-quark particles (known as the mesons).

This still leaves one question, what about the force characteristics of these gluons? Well there is a very special thing about the derivation of the light speed harmonic in the primary gluon, that is the one in the proton ($c^{1/2}/3\pi$, see Chapter 6, box 8) [†]. Specifically, it mathematically represents a force, which constantly acts at 90 degrees, or perpendicular to the object it is orbiting. Thus the force particle must complete a circle around the positron, constantly going at right angles to a line, which constitutes its radius. This is most intriguing since this is exactly the force characteristic required. Moreover this explains logically why the quarks appear to be so confined. As was mentioned

---

[†]Dimensionally, as we have already multiplied the quintessential mass by the number of quanta per unit time [M.T] x [T⁻¹] = [M], when deriving mass of the electron, thereafter we need not repeat this from a dimensional standpoint and the factors of c are treated as pure numbers

before quarks never seem to be seen alone, this is what has been termed quark confinement. In the standard model there is no logical reason given for this quark confinement, other than it seems to be a rule. Now in the new model we can see mathematically why this should be the quark needs to follow the 90 degree rule, which is a direct result of the form of the light speed harmonic we use which is based on the wave equations taken from standard quantum physics (see, Derivation of the electron to proton mass ratio from Schrödinger, Chapter 6, box 8). So now there is a mathematically well-defined and logical reason for this quark confinement.

The force characteristics of the gluons was for some time a mystery. The force characteristics of light, otherwise known as electromagnetism, has been known for some time, in fact since 1864. It was then that a genius scientist called James Clerk Maxwell worked out that light had certain electric and magnetic field characteristics. Essentially in this scheme, if the photon of light is travelling in the x direction then the electric field is oscillating in the y direction and the magnetic field is oscillating in the z direction, the resulting strength of the magnetic and electric field are given in Box 19. It was not until 1954 that two very clever physicists, Yang and Mills, derived the equation for the gluon force field. Now instead of travelling in the x direction the gluon will travel in the z direction forming a closed loop and the electric and magnetic fluxes will be in the x and y direction, this will give its particular closed loop field equation (see Box 19).

Box 19
Photon and Gluon Field Equations

Maxwell's Electromagnetic field equations

$$\nabla \cdot E = 4\pi\rho$$
$$\nabla \cdot B = 0$$

Yang Mills gluon field equation

$$\partial f_{\mu v}/\partial x_v + 2\varepsilon\, (b_v \times f_{\mu v}) + J_\mu = 0$$

where E is the electric field, B is the magnetic field, ρ is the charge density, ∇ is the divergence, f is the strength of the gluon field, ε is the "charge" and J is the related current and b the potential.

There was however one problem with the Yang Mills equation, which has not been resolved, until now. What is the mass of the gluon? Well in this new model, the numerical term $c^{1/2}/3\pi$, multiplied by the mass of the electron would itself would itself represent the combined effective mass of the gluons, which are present in the proton. There are three colours of gluon. These three gluons would bind the three quasi positrons together to form the proton. As for the effective mass of the other seven gluons, this would also depend on their light speed configurations in a similar fashion (as given in Box 18).

Of course the *true fundamental* mass is the quintessential mass of which the electron itself is made up of by means of a spherical harmonic of the speed of light. The mass of the electron then multiplied by a specific harmonic of the speed of light gives the effective (non rest) mass of the gluon. By taking into account these harmonics in the eight gluon masses, (see pages91-92), which in turn explains hundreds of particle masses.

With light (or electromagnetism) that number of (non rest) masses is far, far greater almost giving what is a continuous spectrum over a span of frequencies of over a trillion, trillion, trillion. But in each case both can be shown to be made from the same ephemeral mass quantum. That is the awesome beauty of this model, it explains all of the electromagnetic spectrum and the particle masses on the basis of a single elegant quantum of quintessence.

Chapter 11

Explaining The Neutrino

the most tiny quantity of reality ever imagined by a human.

Frederick Reines

For many years physics experimentalists had been noticing something very strange going on in the subatomic world. It was something that should have caused them quite some consternation. Two of the most important laws of nature were regularly being broken in their particle experiments. One of these most fundamental laws of nature, is that energy is conserved, specifically that the total energy of a system should neither be lost nor gained in any reaction. Another important law was the conservation of the total speed of rotation of a particular body, called the conservation of angular momentum. Typically we see this effect when a spinning ice skater starts of with her hands outstretched and as she brings her arms in she spins faster and faster.

The problem was that whenever the particle known as neutron decayed in to a proton and an electron some energy went missing. The strange thing was, this was a small amount of energy, but it was very random. Equally well some of the angular momentum seemed to be missing. What on earth was going on? Some physicists of the time were willing to let these two laws go, or just ignored the effect. The first one to truly realise what was going on was Wolfgang Pauli, who postulated in 1930 that a hitherto unseen particle was carrying away the energy and angular momentum. Later Enrico Fermi an Italian physicist coined the term neutrino, meaning *little neutral one*.

So small was this particle that it took 25 long years of searching for particle physicists to find it. But eventually it was found by scientists in 1956 and eventually they won the Nobel prize for their discovery. It took another 20 years for scientists to realise that there were three such particles. Originally the Standard Model had placed these particles as having zero rest mass. But in hindsight it was always unlikely that these would have had zero mass, otherwise how could they carry off energy *and* angular momentum.

Recently evidence suggests that the three neutrinos do indeed have a mass each successively heavier than the other neutrino, but that even the heaviest mass is very tiny. This is perhaps another small failing of the Standard Model, which predicted a zero mass. However, the mystery does not stop there; for many years it has been known that the amount of neutrinos reaching us from the sun, was much less than it should be, and for some time nobody knew why. The answer turns out to be even more mysterious. It appears that the smaller neutrinos were spontaneously turning in to the heavier neutrinos during their journey from the sun to the Earth. What could cause this mysterious phenomenon to happen? Well it all turns out to have a perfectly logical explanation.

The fact is that this change to a heavier neutrino nevertheless seems to indicate that neutrinos do have mass. The explanation given by the some scientists, however, varies from being mildly comprehensible to being a little of the wall to say the least, one reasonably sensible argument goes something like this. " *Neutrino flavour eigenstates are not the same as neutrino mass eigenstates, this allows for a calculable probability that an electron neutrino to be detected as either a muon or tau neutrino.*" It is not necessary to try and translate this into understandable language, as there is of course an explanation, which is far more comprehensible once we know what the masses of these three neutrino are. That of course is the other problem physicists know these particles are tiny, but not how much they actually weigh. They merely know the differences between the masses. Here we derive the masses from first principles again we can take our cue from the lepton array.

Box 20
Lepton Neutrino Array[†]

$mv_e = m_q (c^{2½}/4\pi)^{1/2} \pm (\div 3) \approx 1.5 - 4.5 \times 10^{-5}$ eV

$mv_\mu = mv_e (c^{1/3}/\pi) \approx 3.2 - 9.6 \times 10^{-3}$ eV

$mv_\tau = mv_e (c/\pi)^{1/3}.(c/\pi)^{1/9} \approx 0.056 - 0.168$ eV

where c is the number of light speed quanta per unit time, m_q is the quintessential quantum mass, mv_e, mv_μ and mv_τ are the masses of the electron, muon and tau neutrinos respectively.

[†] Dimensionally, as we have already multiplied the quintessential mass by the number of quanta per unit time [M.T] x [T⁻¹] = [M], when deriving mass of the electron, thereafter we need not repeat this from a dimensional standpoint and the factors of c are treated as pure numbers.

Q.E.D. The neutrino masses from the speed of light.

So quite logically the neutrino masses now follow the same pattern as the lepton masses. The caveat is that we cannot achieve five decimal place accuracy with the neutrino masses, as it was possible to do with the lepton masses, because we do not known their magnetic properties. Equally well it is merely prudent to give a range of values. Having said that, by far the most likely masses of the neutrinos are those at the lower end of the range given. Importantly, these levels of prediction are still far, far better than the standard model, which predicted a zero mass.

The most likely values of the equations for the neutrino masses (the lower of the range) suggest that these particles are not only the square root derivatives of the electron but also 1/3 of this. In this case, this suggests the neutrinos are a sort of one dimensional, square root equivalent of the electron. However there is a small possibility they are just the square root derivative of the electron, in which case the masses would be taken as the higher value. † Notwithstanding this, it does appear that the most likely model is the one-third value of the square root of the number of quintessential quanta in the electron divided by three.

So, what exactly does this one-third value suggest? What it suggests is a one-dimensional model of the neutrino, so if you imagine an electron as a ball, which is spinning like a top (on its y axis), then the neutrino would be like an upright ring, which is

---

† There is a possibility they are just the square root derivative of the electron hence the masses would be $mv_e = 4.5 \times 10^{-5}$ eV, $mv_\mu = 9.6 \times 10^{-3}$ eV, and $mv_\tau = 0.16$ eV.

spinning (on its y axis). It still creates a three-dimensional effect, but this would explain why the neutrino is entirely ghostly in its interactions.

What then of their ability to change in to a neutrino with a higher mass. Well now that we know the structure and mass of these particles, this is not difficult to understand. Because the neutrinos are such tiny particles they would be highly relativistic, that is they will have speeds that very nearly approach the speed of light. In other words according to relativity their actual masses will be far, far higher than the actual rest mass. This makes a transition to a higher mass relatively likely.

What then of one of the discoverer's comments about the neutrino:

the most tiny quantity of reality ever imagined by a human.

Frederick Reines

Well the most tiny quantity of reality imagined by this work is the ephemeral quintessential mass and if we calculate how many quintessential quanta there are in the *smallest* neutrino, we get about 3 billion. Only at this level of imagination can one find the one true unifying beauty of quantum physics.

Chapter 12

The Final Piece of the Particle Puzzle

The creation of Physics is the shared heritage of all mankind.

Abdus Salam

The last major piece of the particle puzzle to be found was first postulated by a trio of famous scientists Salam, Weinberg and Glashow, who were later awarded the Nobel Prize in 1979. The actual particles were found in 1983 by teams headed by Rubbia and van der Meer. With unprecedented speed they were also awarded the Nobel Prize, in the very next year in 1984. This alacrity was no doubt because physicists were very happy to have found something, which thoroughly supported the Standard Model of physics. There was no doubt of the genius of the theory, they had managed to predict a particles mass within about 10-20% of its actual mass. But in reality the theory behind the discovery of the particles was a little contradictory, we shall elaborate in due course. Suffice it to say, certainly the particles do exist, but the question is what is their true structure?

    It all started well before 1979 when people were still trying to unify the forces of nature. At the time there appeared to be 4 forces of nature. There was light (otherwise known as electromagnetism), gravity which everybody had known about since the time of Newton. Then there were the newly discovered forces, these held the nucleus of an atom together and were known as the weak and strong nuclear forces. What the trio had done to warrant the 1979 Nobel Prize is to have unified two of the forces of nature, the electromagnetic and the weak nuclear force into the electro-weak force. The problem about physics is that once you have decided that a certain force acts on a body it is then hard to persuade people that it is not a force at all, but a particle. That is because in standard physics the difference between a force and a particle is enormous. A force is something that travels at the speed of light, and has a (non rest) mass that depends on it going at this light speed (or thereabouts). A particle has a rest mass and can, at least theoretically, never achieve light speed, otherwise according to relativity its mass would be infinite. This is where the slight contradiction comes in, what the trio had described and what was actually found, to all intents and purposes, was a particle with a rest mass and a limited speed and not a force.

Perhaps part of their cleverness was to able to think of a way in which a force can appear as a particle. What they did was to use another particle in physics, which is known as the Higgs particle to somehow endow a photon like particle with rest mass. However, the Higgs particle itself remains undiscovered, despite a determined and very costly search for its existence. The other problem with the new particles (termed W and Z) was the particles masses were massive. The W particle for instance was some 80 times the mass of the proton, and therefore some 240 times the mass of the quarks that it purported to describe. This is potentially a problem as the weak force was meant to be exactly that a weak force. Originally invoked to describe the decay of neutron, it was eighty times heavier than the neutron. The third problem was that even when the decay of a neutron to a proton did occur the W particles were not actually seen, but their presence was inferred. In order to actually "see" the particle, very powerful particle accelerators were required.

So the first question to ask is, if it mediates the *weak* nuclear force, why on earth does it appear to have a mass, which is apparently 240 times the mass of the particles that are involved in the *strong* nuclear force? Or in other words, why does a particle decay via a mass which is 240 times bigger than it need be. Well the standard argument is that the energy required is borrowed from the vacuum for a very short period of time, and this is why these spontaneous decay events are relatively rare in the nucleus of an atom. But this argument quickly falls down when we realise that just about all neutrons will decay within about 30 minutes of being outside the nucleus. So it is actually something about the structure of the nucleus that appears to be keeping the neutron stable.

The answer to some of these questions is actually given in chapter 7, (entitled "The Standard Model Remodelled" pg62-65). First of all we need to briefly recap on what the proton is. Well you may recall in Chapter 5 we introduced the concept of the -1/3 fractional negative charge of the electron, which in physics is termed a quasi electron. The mirror particle of this is called the quasi positron with a charge of +1/3 fractional positive charge. So each up quark is a quasi positron, multiplied by a square root light speed harmonic (which in this case represents a gluon). The proton is made of three of these up quarks (see also Chapter 6). Importantly this harmonic is in keeping with the standard equations of quantum electro-dynamics (Q.E.D. see Box 8)

How then do we derive the charges of a particle such as the neutron? In the Standard Model the neutron with no net charge is defined as having three quarks (udd); so doing the calculation shows that the charge is zero (2/3e–1/3e-1/3e = 0). In the new model we no longer need this complex arrangement we simply need to understand the nature of charge. In the standard view, the effects of charge are transmitted by what are known as virtual photons. In truth a virtual photon has never been seen. Also there is another question, which has not been answered by standard physics, why does the electron orbiting an atomic nucleus, as in a hydrogen atom, not disintegrate? All accelerated charges are supposed to give off photons and the electron orbit would thus naturally decay. The standard answer is that the electron orbitals are quantized. However, as with many aspects of quantum physics this is just a reason not to think about the problem - nor to question it.

Well the answer to both questions is straightforward once you understand the *nature of charge*. Recently scientists are beginning to realise that charge is a vortex. To be exact it is a vortex of space-time. Now imagine that you happen to be sitting in such a vortex, it is the space-time around you that is being accelerated so in actual fact you do not feel the acceleration, although you will no doubt be going at some speed. Thus the same applies to the electron orbiting a hydrogen nucleus as it does for an electron in close orbit around a proton to form a neutron.

So now we can formulate a new model. The neutron is now straight-forward. It is a proton with the total charge of +1e, effectively tightly orbited by an electron of charge −1e, which makes the net charge zero (see Fig 2). For proof we can calculate the radius of the orbit of this electron around the neutron and from it the neutrons exact mass (see Figure 2 and Box 21). Once we understand the nature of charge this is entirely possible. It is well known that that the electron going around a hydrogen nucleus would not disintegrate, for the same reason nor does an electron orbiting a proton disintegrate, even in such a confined space.

Having said that, the neutron is not a very stable particle and outside an atom it does disintegrate in about 15 minutes principally into a proton and an electron. This begs the question, why is the neutron so stable when it is *inside* the atom, or even why are atoms so stable? After all, how can a neutral particle such as the neutron hold all those strongly repelling positive charges together? The answer is now direct, the neutron is positive on the inside and effectively negative on the outside and in this way it practically glues the protons in the atom together. Indeed recent experiments confirm this view of the neutron as been more positive in the middle and more negative on the outside.

So first lets compare neutron decay in the Standard Model and in the new model and see which makes more sense. For this purpose we will introduce what is known in physics as a Feynman diagram of decay (See Figure 1). These are just diagrams that help explain the decay process graphically.

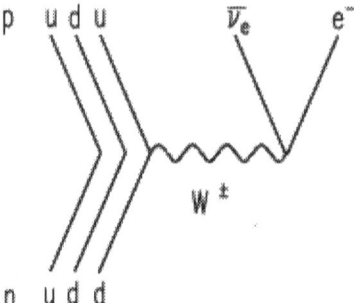

Figure 1, Neutron Decay in the Standard Model.

Lets just talk the reader through what is theoretically happening here in the Standard model. A neutron (n) is made of an up and two down quarks and has zero net charge; (udd = +2/3e-1/3e-1/3e = 0). A down quark changes to an up quark, with the emission of a W particle (with the mass of 80 times that of the proton) and produces a proton (p) with a net charge of +1 (udu = +2/3e+2/3e-1/3e = +1). The W particle then decays into an electron (e⁻) and an anti-electron neutrino ($\overline{V}_e$), (with a combined mass of 160,000 times less than the W particle).

In theory the emission of such a massive particle is possible but only rarely. So why does neutron decay occur so commonly, particularly with decay *outside* the nucleus. Also in the standard model there is no real need for the neutrino to be part of the equation. There is no rotating electron in the standard model hence no conservation of rotation (angular momentum) is necessary. All in all, this a picture of nature that does not exude common sense. Why is there a rigid +2/3 , and minus 1/3 pattern in the quarks. Why is there the spontaneous generation of large particles? There is a more logical model, one that not only explains the decay of the neutron, but also explains its mass and the radius of the neutron. This is perhaps the most important reaction of the Universe, it powers the sun and in turn enables the formation of all the atoms we know. These atoms form the basis of more complex molecules, which enable the formation of life on this planet. The power of the sun is that which sustains this life. Without this reaction and its reverse counterpart nothing would be like the way it is, so to understand this reaction is of paramount importance. The question is can we find a more logical answer, the answer is a resounding yes (see figure 2 and Box 21).

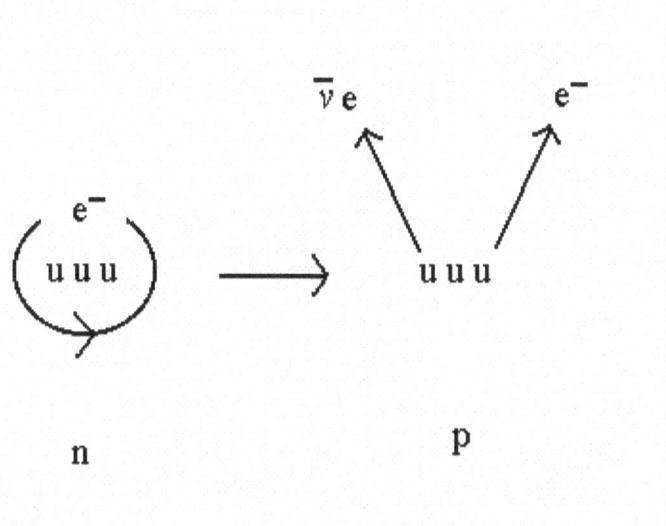

Figure 2, Neutron Decay in the revised Standard Model.

With the revised Standard Model the decay of the neutron is now straight-forward the neutron (n) is a proton (p), composed of three up quarks closely orbited by an electron (e -). It has zero net charge (3x+1/3 = +1-1 = 0). In neutron decay the electron merely becomes unstable in its orbit and flies off at a tangent along with and anti-electron neutrino ($\underline{V}_e$).

For further proof, we can calculate the exact mass and radius of the neutron from first principles. The first question must be, why is the mass of the neutron what it is? The standard model has no accurate answer to this question, except that the down quark weighs a little more than the up quark. In actual fact the neutron actually weighs in at a little more than the proton (at 939.565 MeV). This is equivalent to the mass of the proton plus about two and a half times the mass of an electron. This is where relativity enters the particle puzzle, because the central proton is about 2,000 times heavier than the electron, the electron effectively orbits the proton very closely and thus at great speed. If we calculate this speed it is velocity is approximately 91.863% the speed of light and at this velocity due to well known relativistic effects its mass increases to about 2.53 times the rest mass of the electron, exactly the mass difference between a proton and a neutron (see Box 21).

What about the radius of the orbit of the electron? Well this can be readily explained by just using the calculation for the classic electron radius. If we modify this by using the relativistic mass of the electron, mathematically we come out with exactly the right answer that agrees with the experimental data on the neutron. Moreover, we can also use relativistic quintessence in deriving the equation and it becomes even more elegant (see Box 21).

Box 21
Relativistic Radius of the Electron and the Neutron Radius and Mass

Classical electron radius states that the total mass energy is equal to the potential energy of the elementary charge e spread over a sphere of radius r.

$$m_e c^2 = \frac{1}{4\pi\varepsilon_0} \cdot \frac{e^2}{r}, \quad r = \frac{1}{4\pi\varepsilon_0} \cdot \frac{e^2}{m_e c^2}$$

In relativistic quintessence†: $m_e = m_q \times c^{2\frac{1}{2}}/4\pi^* = h/c^2 \times c^{2\frac{1}{2}}/4\pi^*$

$$r = \frac{*e^2}{\varepsilon_0 \gamma h c^{2\frac{1}{2}}} = 1.113 \times 10^{-15} \text{ m}$$

Or using the standard equation relativistically:

$$r = \frac{1}{4\pi\varepsilon_0} \cdot \frac{e^2}{\gamma m_e c^2} = 1.113 \times 10^{-15} \text{ m}$$

Neutron mass (m_n)

$m_n = m_p + \gamma m_e = 939.565$ MeV

Actual neutron radius (r_n) and mass

$r_n = 1.11 \times 10^{-15}$ m

---

† Dimensionally m_q = [M.T] multiply by n = $c^{2\frac{1}{2}}$, which is the number of light quanta *per unit time*, with the dimensions of [T⁻¹] we get the dimensions [M.T] × [T⁻¹] = [M].

> $m_n = 939.565\ MeV$

where ε_0 is the electric constant, c is the speed of light, e^2 is the unit of charge in coulombs. m_e is the mass of the electron ε denotes the standard electron magnetic moment to Bohr magneton ratio, * = (1 + 2ε). and $\gamma = 2.53$, where $\gamma = 1/(1 - v^2/c^2)^{1/2}$

This means effectively that if we use the relativistic quintessence model or the relativistically modified classical methods for calculating the radius of the electron we get the right answer. Ironically the classical method has been ignored by many.

What most people are aware of is the quantum method for deriving the radius of the electron around the hydrogen atom. Indeed this quantum method is well recognised and accurate when considering the electron *as a wave* (and has been derived from first principles in Book 1 of this series). The relativistic quintessence model or the relativistically modified classic method for calculating the electron radius, now appears more pertinent than ever, precisely when you are considering the electron *as a solid charge*. We can also use the relativistic quintessential quantum for deriving the radius, which is both elegant and can be used to derive the exact mass and radius of the neutron from first principles.

Additionally, we no longer need a massive particle with a mass of in excess of eighty times the mass of the proton, to be spontaneously generated to explain neutron decay. The new model (see Fig 2) is entirely more logical. The presence of a tiny, tiny neutrino is easily explained by the need to conserve the rotation (angular momentum) of the rotating electron.

Better still, there is no need to create the weak nuclear force at all, in this type of reaction. The force we already have, the electric force (technically the electromagnetic force) is all that is needed (Q.E.D see Box 21). One merely needs to look at derivation of the classical radius of the electron, which we re-derive from first principles by taking the relativistic mass of the electron (see Box21). So we can use standard electric force to understand the "fluid" radius of the electron and in turn derive the radius of the neutron. The important thing is that the electron radius around a hydrogen atom, is derived by viewing the electron as a (gaseous) wave (see Book 1). When the electron is considered as a "fluid" charge, as opposed to a "gaseous" wave, the radius comes out a lot smaller and hence corresponds to the radius of the (relativistic) electron charge radius.

Under these circumstances we can not only use this equation to derive the radius of the neutron (see Box 21) but we can also use it for deriving the radius of the proton. We do not need to alter the electron mass, as this is the fundamental harmonic mass, we merely need to calculate the relativistic velocity of the specific charge elements (in the case of a proton these are positrons) to arrive at the correct radius. So using the relativistic equations shows these elements in the proton are travelling at 95.196% of the velocity of light

So once one understands the true nature of charge, everything drops out from first principles. Now we can understand what a proton and a neutron actually is, we can remodel the decay of the neutron, without needing these new W particles, which have in any case not been seen in these sorts of decays.

Crucially it turns out that the standard equations for the electric charge are sufficient to account for this manifestation of the weak nuclear force. Probably the most important fundamental reaction in the Universe can now be explained using the electric force and the single exquisite quintessential quantum in a magnificently designed and logical way.

※

This is not to say that the W and the Z particles do not exist, they clearly do, but they are actually only seen in other types of decay processes. The internal structure of these W and Z particles needs to be seen to understand why they play this other role in particle physics.

The questions left, are what actually are W and Z, and are they particles or forces. Well we should stick with the experimental data and call them particles. In this case there is no need to create a fourth force, the weak nuclear force. So what sort of particles are they and how is their mass generated? In physics these particles are called intermediate vector bosons and actually this describes their structure and function very nicely. Because the W particle, or intermediate vector boson's function is to mediate between particles of a different type (the higher order leptons and the quarks), then the structure should reflect the function. So the mathematical structure of these particles allow them to decay into a quark like structure, one that we see in the proton for instance $(c^{1/2}/3\pi)$ or a mathematical structure which we see in the higher order leptons, for instance $(c^{1/3}/\pi)$ (see Box 22).

Box 22
Intermediate Vector boson mass ($W^{\pm}$) and Decay Products‡

$$W^+ = {}^*m_e . (c/2\pi^6) = 80.43 \, GeV$$

Actual W^+ mass

$$W^+ = 80.43 \, GeV$$

Decay in to baryonic matter

$$W^+ = m_e . (c/2\pi^6) \rightleftharpoons \tfrac{1}{2}(c^{1/6}/\pi)^6$$

$$W^+ \rightleftharpoons m_e . \tfrac{1}{2}(c^{1/6}/\pi)^3 . (c^{1/6}/\pi)^3 \Rightarrow m_e . \tfrac{1}{2}(c^{1/2}/3\pi) . (c^{1/2}/3\pi)$$

Or decay into Leptonic matter

$$W^+ \rightleftharpoons \tfrac{1}{2}(c^{1/6}/\pi)^6 \rightleftharpoons m_e . \tfrac{1}{2}(c^{1/6}/\pi)^2 . (c^{1/6}/\pi)^2 . (c^{1/6}/\pi)^2$$

$$W^+ \Rightarrow m_e . \tfrac{1}{2}(c^{1/3}/\pi) . (c^{1/3}/\pi) . (c^{1/3}/\pi)$$

c is the number (n) of speed of light quanta, ε is the standard proton magnetic moment to Bohr magneton ratio, *= (1 + 2πε) and m_e is the mass of the electron.

---

‡ Dimensionally, as we have already multiplied the quintessential mass by the number of quanta per unit time [M.T] x [T⁻¹] = [M], when deriving mass of the electron, thereafter we need not repeat this from a dimensional standpoint and the factors of c are treated as pure numbers

Q.E.D. The mass of the intermediate vector boson from the speed of light.

Now we have seen the mathematical structure of the intermediate vector boson, we can predict pretty much exactly how particles of a higher mass are going to decay. Until now this was determined experimentally, but from here on in we can progress from the Feynman line diagrams (see Figures 1 & 2) to see how a particle decays, to the use of the actual mathematical light speed structures of particles to plot particle decay. Moreover it becomes absolutely clear why these bosons are not seen in decays to the electron, but are involved in those involving the second and third level leptons (the muon and the tau).

In order to corroborate the quark structures these structures need to explain in detail the pattern of decay of the quarks themselves. We are thus required to explain particle decay from first principles including the structures, which are formed in these decays, such as the mediator of the electro weak force (the Intermediate Vector Boson). To see if this is possible the decay of the truth (t, top quark) will be examined. In very high-energy experiments a truth and an anti-truth quark are produced The decay of each merely mirrors the other in most cases, and we will just follow the decay of the truth.

According to experiment the truth quark most commonly splits into two particles, the beauty quark (b, bottom quark) and the intermediate vector boson (W^+). Now we can follow these decays by seeing their mathematical, structure, all based on a harmonic of the speed of light multiplied by the mass of the electron.

Box 23
Decay Modes of the Truth Quark (t)‡

$$ch = m_e \, (c^{1/3}/\pi).(c^{1/4}/3\pi)$$
↗
$$b = m_e \, (c^{1/3}/\pi^{1/2}).(c^{1/4}/3\pi^{1/2})$$
↗
⇒⇒ $$t = m_e \, (c/\pi)^{1/3}.(c/\pi)^{1/4}.(c/\pi)^{1/9}$$
↘
$$m_e \cdot \tfrac{1}{2}(c^{1/3}/\pi^2)^3 + \gamma$$
↘
$$W = m_e \cdot (c/2\pi^{\,6})$$

↙ or ↘

$m_e \cdot \tfrac{1}{2}(c^{1/6}/\pi)^3 \cdot (c^{1/6}/\pi)^3$ $m_e \cdot \tfrac{1}{2}(c^{1/6}/\pi)^2 \cdot (c^{1/6}/\pi)^2 \cdot (c^{1/6}/\pi)^2$

↘   ↘

$u = m_e \cdot (c^{1/2}/9\pi) + \gamma$ $\mu = m_e \cdot (c^{1/3}/\pi) + \gamma$
 +   +
$u = m_e \cdot (c^{1/2}/9\pi) + \gamma$ $v_\mu = m v_e \, (c^{1/3}/\pi)$

where c is the number (n) of light speed quanta, m_e is the mass of the electron μ is the muon, v_μ is the muon neutrino. u, ch, b and t are the up, charm, beauty and truth quarks, W is the intermediate vector boson, γ represents a photon Particles not in bold are interim particles. For *accurate* particle masses please see separate Boxes 4,7,11,13,15,16,17,18 and 22.

---

† Dimensionally, as we have already multiplied the quintessential mass by the number of quanta per unit time [M.T] x [T⁻¹] = [M], when deriving mass of the electron, thereafter we need not repeat this from a dimensional standpoint and the factors of *c* are treated as pure numbers

So if we examine these mathematical structures we can actually almost see the very decay process in action. This supersedes those very clever Feynman diagrams, which can only give a graphical, but not mathematical account of what is actually happening. For instance, if we look at the beauty quark we can immediately see, which parts of the top quark contribute to its structure (see Box 23). Similarly the beauty quark immediately decays in to a charm quark with a light speed harmonic structure, which is merely beauty divided by pi (π), with the release of a specific photon.

Because of the structure of these particles, whenever we move from a higher tier of particle to a lower tier of particle we generate an intermediate vector boson. In the new model therefore, when truth breaks up in to a lower tier quark, the beauty quark, it produces an intermediate vector boson. This does not apply to the decay of beauty to charm because they are on the same tier. Indeed this new Model is a far better reflection of what actually happens in real experiments than the Standard model. The mathematical structure of the intermediate vector boson immediately shows how it can decay either into a quark like structure or a lepton like structure, in this case a muon and its respective neutrino.

So it would appear that the intermediate vector boson is not a force at all, it is itself a particle, which acts as an intermediate particle between the tiers (or generations) of particles. Like the quarks it is merely an electron multiplied by a light speed harmonic. This is however a special light speed harmonic, one which can act as a bridge between the lepton type particles, and the baryon type particles, like a quark. Indeed the mass of the particle is far higher than the mass of the other quarks themselves. And when it decays it most frequently decays in to quark like structures. Hence, it's mathematical structure suggests it is probably best thought as derived from a gluon - a ninth gluon. Indeed this means we have *united the W and Z particles, specifically the weak nuclear force, not with the electric force, but with strong nuclear force.* Because of the magnitude and structure of these intermediate vector particles, this is exactly where it should belong. So now we merely have three forces of nature, the electromagnetic, gravity and the strong nuclear force.

What then of the supposed weak nuclear force and the decay of the neutron into a proton. Well if you look at this reaction it principally involves a change of charge and not of mass and we can readily invoke the electric force to mediate this reaction (see Box 21). In fact the neutron is merely a proton with an electron in close orbit around it (see Figure 2) Indeed it can be shown, using conventional physics, that this reaction is part of the electromagnetic force -and a very important part too.

One more major particle mystery remains to be solved. Why is there, apparently, matter but little or no antimatter in the Universe? According to common sense these should be present in equal amounts in the Universe. Well in actual fact they *are* present in equal amounts. The electrons orbit on the outside of the atom and its anti particle, the positron (along with gluons) makes up the proton, and is locked on the inside of the atom in the nucleus. Equally well it turns out the neutron is in fact a proton made of a positron (along with gluons) tightly orbited by an electron and we can use standard physics to show this (see Figure 2 and Box21) As a result matter and anti matter are effectively present in equal amounts. So how can we be so sure of this neutron structure? Ironically the use of standard physics can prove this (see Box 21). It is possible to calculate the exact mass and radius of the neutron form these observations from first principles (see Box 21). The Standard model or indeed string theory cannot come close to predicting these properties. The other question is why should the particles be this way round? Well this is not directly to do with the particles themselves, but the way in which the universe itself is formed, and you'll have to read the next book (Book 3) to find that out.

Finally, what does this now tell us of that particle that is supposed to endow all other particles with mass, the famous Higgs particle? Everybody is spending billions upon billions searching for this particle, as it is the very last and most important piece of the particle puzzle. Well we have already found the fundamental particle that endows mass - that is the new quintessence. Indeed the beauty of the new model is that you only need this one quantum and the speed of light to explain the masses of all the known particles. Hence we can relatively confidently predict that *there is no Higgs particle* .[†]

What can be deduced however, is that like the Higgs, there is a particle which is present as a vacuum field, and endows particles mass as they pass through the vacuum. Indeed this particle makes up the very

---

[†] You might find that there is a pentaquark or other such particle, but a spin-less, charge-less Higgs particle between 115-175 Gev, is extremely unlikely to be found.

fabric of space-time. That particle is clearly the one and the same new quintessence quantum, which also makes up everything else in the physical Universe.

That completes the entire particle physics puzzle and the forces of Nature, which act between them. All the particles and the forces between them are made from the primary quantum of the Universe, the ephemeral quintessential quantum, multiplied by their respective light speed harmonic. Crucially there is only the need for *one* fundamental quantum particle, the new quintessence, from which the rest is exquisitely elegantly constructed.

Glossary of Terms

From the Alpha to the Omega

We have in this book derived the fundamental constants of particle physics using the quintessential quantum from direct first principles in an entirely elegant way. This has been done solely from, the electric constant, Planck's constant and the velocity of light.

Here we list those elements of particle physics, which it has been possible to derive from first principles:

Alpha, *symbol α* :
Fine structure constant. effectively the ratio of the speed of an electron orbiting a hydrogen atom to the speed of light, (see Box 6).

Alpha Particle, *symbol α particle* :
Decay product of the nucleus, later found to be the same as a helium nucleus, consisting of two protons (see Box 7) and two neutrons (see Box 21).

Baryon :
Subatomic particle group, of varying masses made of three quarks (see Box 11).

Beta Particle, *symbol β particle* :
Originally named after beta decay (see Figure 2, Box 21), later found to be the same as an electron (see Box 4).

Constancy of the speed of light, *symbol c* :
Speed of light, fundamental constant. Light speed harmonics form the fundamental basis of the difference between masses of all the known subatomic particles (see Boxes 1-23).

Coulomb, *symbol C* :
Electric charge unit. Charge of the electron fundamental charge (see Box 5).

Delta Particle, *symbol Δ particle* :
Specific baryonic particle, made from three constituent quarks. (see Box 11).

Electron, *symbol e* :
Fundamental mass of the electron, leads to the mass of the other heavier particles, (see Box 4, 7-23).

Frequency, *symbol f* :
Fundamental characteristic of all known particles and forces (see Eq. 1), endowing particles their wavelike properties.

Gamma, *symbol γ* :
Symbol commonly used for a (high energy) photon, often resulting from particle decays (see Box 23)

h, Symbol for Planck's constant.
Fundamental unit of energy or action. Essential to quantum physics and for the construct of a fundamental quantum of mass (see Box 1).

Higgs particle:
Hypothetical particle, thought to endow mass to the particles of the Universe. As yet unfound and unlikely to exist as predicted.

Imaginary number, *symbol i* :
The square root of −1. Useful mathematical tool, often used in the derivation of the Schrödinger wave equation (see Book 1, Box 13 and Book 2, Box 8)

j, also a symbol for an Imaginary number:
The square root of −1. Useful mathematical tool, often used in the derivation of the Schrödinger wave equation (see Book 1, Box 13 and Book 2, Box 8).

Kaon, *symbol K*:
Specific Meson particle made of two quarks (see Box 12).

Light :
Light is a form of electromagnetic radiation. Light speed harmonics form the fundamental basis of the difference between masses of all the known subatomic particles (see Boxes 1-23).

Meson:
Subatomic particle group of varying masses made from a pair of quarks (see Box 12).

Muon: *symbol μ* :
Second generation lepton particle (see Box 13).

Neutron, *symbol n* :
Fundamental neutral baryonic particle made from three quarks, often also called a nucleon as it sits in the nucleus of an atom (see Box 21 and Figure 2).

Omega, *symbol Ω:*
Specific baryonic particle made of three quarks. (see Box 11)

Proton, *symbol p :*
Fundamental positively charged baryonic particle made from three quarks often also called a nucleon, as it sits in the nucleus of an atom (see Box 7 and 11).

Quarks, *symbols u, d, s, c, b, t.*
Up, down, strange, charm, beauty and truth. Fundamental constituents of baryonic and meson matter of varying masses previously of unknown origin, now deduced from the quintessence mass. (see Boxes, 11, 12, 15, 16, 17, 19).

Quintessence Mass, *symbol m_q:*
Fundamental quantum mass unit of the Universe, endows all particles and forces with mass (see Box 1-23).

Rho, *symbol ρ*
Specific meson consisting of two quarks (see Box 12)

Sigma, *symbol Σ :*
Specific baryon consisting of three quarks (see Box 11)

Tau, *symbol τ :*
Third generation lepton particle (see Box 14)

Upsilon, *symbol γ:*
Heavy meson made of two beauty quarks (see Box 16).

v_e Symbol for the electron neutrino,
Particle with a previously unknown mass, now deduced (see Box 20).

W, symbol for the intermediate vector boson.
Thought to be the charged mediator of the electro-weak force (see Box 22,23)

Xi particle, *symbol* Ξ
Baryonic particle made for three quarks (see Box 11)

Y, Why?
A question not asked frequently enough (see Box 1-23).

Z, symbol for the intermediate vector boson.
Thought to be the neutral mediator of the electro-weak force, related to W (see Box 22).

Technical Notes.

1). *Frequency*
Common questions arise from this straightforward *a priori* assertion, $f = n$, *the frequency is equivalent to the number of quanta, per unit time,* these can readily be answered.

a.) How can a number have the dimensions of frequency? Well it is actually the number of quanta *per unit time*, so it will have the dimensions of frequency, specifically [T^{-1}].

b.) Another question is what *are* the units of time? Well the units of Planck's constant h are given in Joule seconds (J s). Hence the unit of time of the frequency must be given in seconds (s^{-1}).

c.) A much more philosophical question arises, does it matter which units of time you use? The answer is, *no* it does not matter which unit of time you use, provided you are consistent, you get the very same answers.

This is where some people have some difficulty. The fact remains that time *elapsed* is *not* the same as *units* of time. Time can elapse, in this case the more time that time elapses the smaller the energy component of the minimum quantum gets as h, which consists of energy multiplied by time, is a constant. Visa versa the less time that elapses the greater the energy component of the minimum quantum is.

Nevertheless, when we change *units* we cannot do this in isolation, for the equation must balance. For example if we change from S.I. units to cgs units, then not only does the meter change to centimetres, but kilograms change to grams and energy changes to ergs. To get the equivalent answer in Joules we have to convert ergs back to Joules and the same answer emerges, provided we use the same actual quantities, whatever the units. The important thing is because we have changed one unit we also have to change other units, we cannot change units in isolation. Indeed the equation $E=mc^2$, must hold.

This aspect is very important, so it is worth staying with the explanation. Lets now change the time unit and see what happens. The fact is if we are using Joules then to balance the equation then if we increase the time unit we would have to increase the either length unit, or the mass unit to balance the units. So what happens when we increase the time unit. Lets say we increase the units from seconds to minutes. If we take the time elapsed for example as 1 second. Then 1/60th of a minute will have elapsed and the energy component of the quantum h, will as before appear to rise by 60. But remembering that the length must also change means that the unit of length goes up by 60 also, as length is a component dimension of energy [$ML^2\ T^{-2}$] when the unit length component goes up the energy decreases by 60. So in fact if you change the unit of time T, you have to increase the length L dimension. The two changes balance and you get the same answer h, for any new unit of time.

We can do exactly the same with time and change the unit of mass, in this case to balance the units, mass needs to go up whatever the time units went up, but squared to keep the equation balanced. It is not necessary to go through the whole explanation again to see that the two changes balance an you get the same answer h, for any new unit of time.

The important thing is for every unit change the equation $E = hf$ *is the same for all time units used.* The main thing to remember when working this all out, is to remember time *elapsed is not the same as units of time.*

This is the absolute conceptual beauty of these observations, so whatever time unit you use h is effectively the same , the frequency f is therefore the same, the number of quanta per unit time n is the same, and m_q is the same.

To prove this we just need to work out for example m_q in S.I units and then in cgs and see that we get exactly the same answer.

Box 19
Lets do S.I. units first

$m_q = h/c^2$

$h = 6.626 \times 10^{-34}$ J s

$c = 2.9979 \times 10^8$ m/s

$m_q = 7.373 \times 10^{-51}$ kg s

Then lets do it in cgs

$m_q = h/c^2$

$h = 6.626 \times 10^{-27}$ erg s

$c = 2.9979 \times 10^{10}$ cm/s

$m_q = 7.373 \times 10^{-48}$ g s $= 7.373 \times 10^{-51}$ kg s

Q.E.D.

It would appear that the Universe is trying to introduce a beautiful new concept, not only is space-time interlinked but energy and space-time are interlinked. We should have guessed this from $E=mc^2$. But now that the science is telling us that there is energy inherent in apparently empty space that's evidence enough to support it. This takes us to the next common question.

2). Dimensionalty

The conventional formula for the Planck mass is dimensionally constrained to give a Planck mass value, with the dimensions of M which is difficult to use in string theory.[7,8] The quintessential mass has the dimensions [M][T], which when multiplied by the frequency with the dimension [T^{-1}], represented by the number of quanta per unit time we resolve the dimension back to those of M. From this result, it is also clear that dimensionally, the number of quintessences (n) is directly equivalent to the frequency, in units of sec^{-1}. Therefore the dimensions of the effective mass of the system, $m = m_q.n$, are entirely consistent with the dimensions of matter.

$$M = [M][T][T^{-1}]$$

These dimensions are also compatible with those of The Planck energy itself whose dimensions are [E][T]such that from the equation $E = hf$.

$$E = [E][T][T^{-1}]$$

It is quite clear that while the Planck energy is the key to understanding energy relations at the quantum level it is equally important to have a fundamental mass, which conforms to the Planck scale.

3). Volume of all the Oceans

Volume of all the oceans = 1.37 billion km^3 =1.37 billion, billion m^3 = 1.37 billion, billion, billion mm^3. So, one tenth of a billionth of a billionth of a billionth = 0.137 mm^3 = volume of a tiny droplet of mist.

References.

1. Polkinghorne, J.C. The quantum World; Longman, Edinburgh (1984).

2. Hay, T and Walters, P. The Quantum Universe; Cambridge University Press, Cambridge (1987).

3. Byron Jr. F.W. Fuller, R.W. Mathematics of Classical and Quantum Physics. Dover Publications Inc. New York (1992).

4. Wojciechowski, A.P. Derivation of a Quantum Based Cosmological Energy Equivalence formula. www.wwk.org.uk/articles/arxiv.999.pdf.

5. Worsley, A.P. Twist, P.J. Generation of a force on a rotating body such as a Superconductor.
Patents and Designs Journal, **5841,** 1613. (2001).

6. Planck, M. "Zur Theorie des Gesetzes der Energieverteilung im Normalspektrum." *Verhandl. Deutsch. phys. Ges.* **2**, 237, 1900.

7. Planck, M, "Über das Gesetz der Energieverteilung in Normalspektrum." *Annalen der. Physik* **4**, 553, 1901.

8. Polchinski, J.. String Theory. Cambridge University Press, Cambridge (1998).

9. Witten, E. Duality space-time and quantum mechanics. *Phys. Today* **50,** 28-33 (1997).

10, Green, M.B., Schwarz, J.H.& Witten, E.. Superstring Theory, Cambridge University Press, Cambridge (1987).

11. Ellis, J. *et al.* Search for violations of quantum mechanics. *Nuc Phys B* **241,** 381-405 (1984).

12. Ellis, J. *et al.* String theory modifies quantum mechanics. *Phys Lett B* **293,** 37-48 (1992).

13. Marie Curie Skłodowska,. Les Rayon Alpha. *Comptes Rendu.* **130,** 76 (1900)

14. Seiden, A. Particle Physics, A comprehensive Introduction Pearson, Addison Wesley, San Francisco (2005).

15. de Broglie, L. Waves and Quanta. *Comptes Rendus* **177,** 507-510 (1923).

16. Einstein A. Zur Electrodynamik bewegter Korper. *Annalen der Physik, 17, 1905*

17. French, A.P. Special Relativity. Chapman and Hall 1968.

www.ingramcontent.com/pod-product-compliance
Lightning Source LLC
Chambersburg PA
CBHW030802180526
45163CB00003B/1134